Surreal

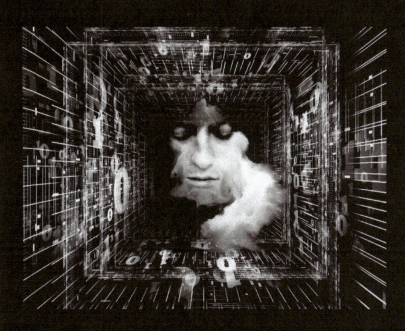

Technology

Scott Tilley

Surreal Technology

Cover design © Scott Tilley

Cover photograph © agsandrew/Shutterstock

Published by CTS Press

CTS
Press
www.CTS.today

An imprint of Precious Publishing, LLC

Precious Publishing
www.PreciousPublishing.biz

ISBN-13: 978-0-9979456-4-5

ISBN-13: 978-0-9979456-5-2 (ebook)

TABLE OF CONTENTS

DEDICATION

To Gene Roddenberry – the creative genius behind "Star Trek"

PREFACE

Merriam-Webster's word of the year for 2016 was "surreal." According to the online dictionary, surreal is "used to express a reaction to something shocking or surprising." Surreal is also a good word to describe some of the key events in the technology world that took place in 2016. In particular, advances in artificial intelligence and big data, and continuing problems with cybersecurity.

"Surreal" could also be used to describe the fictional but prescient technology that appeared in the original "Star Trek" television show (ST:TOS) in the 1960s. Hand-held communicators, faster-than-light travel, and laser weapons were all born from the imaginations of the show's writers. But now, 50 years later, much of that fiction has become fact. In 2016 I gave several public lectures on "Star Trek at 50" that addressed issues related to communication, transportation, and civilization that were first introduced in ST:TOS. I based the lecture on the columns about "Star Trek" included here.

I hope you find this collection of my "Technology Today" columns from 2016 interesting. The column appears each week in the business section of the *Florida Today* (Gannett) newspaper. Any errors or omissions in the book are mine alone.

I can be reached via email at TechnologyToday@srtilley.com. You can follow my column @TechTodayColumn on Twitter. I'm also on Facebook: www.facebook.com/stilley.writer. Learn more about my other books at www.amazon.com/author/stilley.

Scott Tilley
Melbourne, FL
May 21, 2017

ACKNOWLEDGMENTS

In 2016 I launched two new initiatives: *Big Data Florida* (www.BigDataFlorida.org) and the *Center for Technology & Society* (www.CTS.today). Both have grown into successful ventures, but it would not have been possible without the many people who helped along the way. Particular thanks to Daryl, Tauhida, Karl, and Cliff for their support. Some of the topics covered in Big Data Florida meetings and CTS events are included as columns in this book.

Thanks to the home zoo (Hopper, Ripley, Watson) for keeping me active and away from work by forcing me to take play breaks.

Lastly, as always, thanks to all the loyal followers of my Technology Today column. I appreciate your correspondence and feedback. Without such great readers, this wonderful compilation would not exist.

timeanddate.com

Calendar for year 2016 (United States)

January

S	M	T	W	T	F	S
					1	2
3	4	5	6	7	8	9
10	11	12	13	14	15	16
17	18	19	20	21	22	23
24	25	26	27	28	29	30
31						

○:2 ●:9 ◐:16 ○:23 ◑:31

February

S	M	T	W	T	F	S
1	2	3	4	5	6	
7	8	9	10	11	12	13
14	15	16	17	18	19	20
21	22	23	24	25	26	27
28	29					

●:8 ◐:15 ○:22

March

S	M	T	W	T	F	S
		1	2	3	4	5
6	7	8	9	10	11	12
13	14	15	16	17	18	19
20	21	22	23	24	25	26
27	28	29	30	31		

○:1 ●:8 ◐:15 ○:23 ◑:31

April

S	M	T	W	T	F	S
					1	2
3	4	5	6	7	8	9
10	11	12	13	14	15	16
17	18	19	20	21	22	23
24	25	26	27	28	29	30

●:7 ◐:14 ○:22 ◑:29

May

S	M	T	W	T	F	S
1	2	3	4	5	6	7
8	9	10	11	12	13	14
15	16	17	18	19	20	21
22	23	24	25	26	27	28
29	30	31				

●:6 ◐:13 ○:21 ◑:29

June

S	M	T	W	T	F	S
			1	2	3	4
5	6	7	8	9	10	11
12	13	14	15	16	17	18
19	20	21	22	23	24	25
26	27	28	29	30		

●:4 ◐:12 ○:20 ◑:27

July

S	M	T	W	T	F	S
					1	2
3	4	5	6	7	8	9
10	11	12	13	14	15	16
17	18	19	20	21	22	23
24	25	26	27	28	29	30
31						

●:4 ◐:11 ○:19 ◑:26

August

S	M	T	W	T	F	S
	1	2	3	4	5	6
7	8	9	10	11	12	13
14	15	16	17	18	19	20
21	22	23	24	25	26	27
28	29	30	31			

●:2 ◐:10 ○:18 ◑:24

September

S	M	T	W	T	F	S
				1	2	3
4	5	6	7	8	9	10
11	12	13	14	15	16	17
18	19	20	21	22	23	24
25	26	27	28	29	30	

●:1 ◐:9 ○:16 ◑:23 ●:30

October

S	M	T	W	T	F	S
						1
2	3	4	5	6	7	8
9	10	11	12	13	14	15
16	17	18	19	20	21	22
23	24	25	26	27	28	29
30	31					

◐:9 ○:16 ◑:22 ●:30

November

S	M	T	W	T	F	S
		1	2	3	4	5
6	7	8	9	10	11	12
13	14	15	16	17	18	19
20	21	22	23	24	25	26
27	28	29	30			

◐:7 ○:14 ◑:21 ●:29

December

S	M	T	W	T	F	S
				1	2	3
4	5	6	7	8	9	10
11	12	13	14	15	16	17
18	19	20	21	22	23	24
25	26	27	28	29	30	31

◐:7 ○:13 ◑:20 ●:29

Jan 1	New Year's Day	May 30	Memorial Day
Jan 18	Martin Luther King Day	Jun 19	Father's Day
Feb 14	Valentine's Day	Jul 4	Independence Day
Feb 15	Presidents' Day	Sep 5	Labor Day
Mar 27	Easter Sunday	Oct 10	Columbus Day
Apr 13	Thomas Jefferson's Birthday	Oct 31	Halloween
May 8	Mother's Day	Nov 8	Election Day

Nov 11	Veterans Day
Nov 24	Thanksgiving Day
Dec 24	Christmas Eve
Dec 25	Christmas Day
Dec 26	'Christmas Day' observed
Dec 31	New Year's Eve

LOOKING AHEAD TO 2016

Three things that will NOT happen this year

January 1, 2016

Gazing into a crystal ball to predict the future is a mug's game in the technology world. What you expect to happen often doesn't come to pass, and what does happen is often totally unexpected. So instead of trying to forecast likely developments in 2016, let's take a look at three things that I posit will NOT happen – hype notwithstanding.

Print books will die. This is something we've been told ever since the first e-book reader was released over 15 years ago. It's true that sales of print books dipped for a while, but whether this was due to e-books or a general decline in readership is unclear. I have no doubt that the future of books remains digital, but to paraphrase Mark Twain, the death of the print book has been much exaggerated.

Sales of e-books have declined somewhat, but this may be due to pricing more than preference. Traditional publishers are doing whatever they can to hold onto their market, which is giving independent book publishers a unique opportunity to grow their readership. For many traditionalists, there's just no substitute for the experience of reading a book in print: it's tactile, efficient, and never requires recharging.

Virtual reality (VR) will go mainstream. Large companies like Facebook are investing literally billions of dollars in VR hardware and software. Their hope is that consumers will embrace the immersive experience of artificial sight and sound when exploring digital worlds, such as those found in video games. But we've seen this happen before: the technology gets better, but the general population fails to adopt it.

Some people dislike the disconcerting feeling of lagging movement in the VR glasses, which can make you feel like you're sea sick. The biggest barrier to success for VR may be the ridiculous goggles you have to wear: they're uncomfortable, make you look like a dork, and cut out real-world social interaction. Can you seriously imagine the whole family sitting around the TV wearing them?

<u>Education will be radically changed</u>. Education is one of the toughest nuts to crack. Over the years there have been many attempts at educational reform through the judicious use of technology. Some of these efforts have paid off, but many have not. Simply throwing more money at the problem isn't the answer either – just ask Mark Zuckerberg about his $100 million donation to the Newark, NJ school system.

But failure doesn't mean we should stop trying. Technology can make a difference, and there are few areas like education where a positive impact would have such far-reaching effects on society. I'm confident we'll eventually get it right.

# # #

FLEXIBLE DISPLAYS

Roll-up TVs are just the start of the smart material revolution

January 8, 2016

New TVs are always highlights of the sprawling Consumer Electronics Show in Las Vegas. Every year the TVs get simultaneously larger and thinner. The images and videos used to demonstrate the television's clarity look better than what's on offer in the real world. I've seen some gargantuan models that are so big you'd need to take your front door off to get the TV into your house.

Most current TVs use LCDs to produce the picture you see. OLED TVs are slowly becoming available, and while their displays are undeniably superior to a standard LCD, they are quite expensive. But even discounting the price, the TV is still essentially the same: a big, rigid glass rectangle that sits on a stand or is mounted to a wall.

At CES 2016, LG demonstrated what may truly be a game changer for TVs of the future: flexible displays. These prototypes are still HD quality, but the display can be rolled up like a newspaper. Imagine how such a display would change how we experience television. There would be no need to permanently mount those giant 65" TVs on the wall when you can just have the TV roll up and down like a blind as needed.

If you've been on some of the newer cruise ships and seen the artificial portals used for interior cabins, you've had a glimpse of the future. The portals display changing scenery from outside the ship, giving the occupants the illusion that they have a large window onto the world. It's quite effective.

Now imagine a future generation of these flexible displays that are incorporated into smart materials, like the walls of your home, or the paint on the walls. The wall could change function at the flip of a switch: from nothing to perhaps a high-definition version of a famous painting to a fully interactive video display. The size of the active display space on the wall could be changed at will. You could have these display-enabled spaces on as many walls (or ceilings, floors, windows, etc.) as you liked.

Consider as well how flexible displays and smart materials would change how you use computers. It would be possible to turn any surface into an iPad-like device. All your data would be accessible from the cloud. In effect, you would no longer need to carry any computer around with you (except for perhaps a smartphone), because computing capability would be ubiquitous.

Of course, nothing says these displays have to be one-way. It would be trivially easy to embed cameras and sensors into them. Then you'd become your own Truman show.

# # #

BACKUPS

Do it or lose it

January 15, 2016

Do you have a backup strategy in place? Do you use it consistently? Have you verified that it works? If you said "no" to any of these questions, it's only a matter of time before a catastrophic event occurs and you lose your valuable data. At that point, it's too late.

But first, let's back up a bit. Figuratively, not literally.

The holidays have been over for two weeks. I suspect that many people still haven't sorted through the hundreds of digital photos and movies taken on smartphones. If you haven't already done so, take the time to transfer the files from your phone to your computer (or tablet), so that they can be filed properly – typically using something like Apple's "Photos" app.

The upside of smartphones and digital cameras is that it's become very easy to take pictures of everything. With no actual film to worry about, there's no need to worry about wasting a shot: if you don't like it, delete it. The downside of smartphones and digital cameras is that this ease of use has led to a tremendous increase in the number of data files to save. I have literally dozens of folders on my computer named "To Be Sorted" that are full of older pictures.

This increased photo volume makes it all the more important to keep copies stored safely somewhere. There's no negative in the envelope anymore; there's only data. And once lost, it's never coming back.

So, let's backup a bit. Literally, this time.

My approach to backups might be called the "belt, suspender, and Velcro" technique. I'm a bit paranoid over losing data because I've been burned the past. Just like the family who installs a home alarm system after their house has been burgled, I've learned the hard way to keep backups up-to-date at all times.

I use a product called CrashPlan that constantly runs in the background on my Mac, copying files from the computer to the cloud. If I lose a local file, I can usually retrieve it online – no matter where I am. It also means I can restore all of my data from scratch if needed, even if I had to replace the machine.

I also backup locally using Apple's TimeMachine program. This periodically copies data over Wi-Fi to a drive connected to my router. Backups are accessed using a slick time-elapse interface.

Lastly, I occasionally make copies of particularly important data files and store then on a portable hard drive, which I keep outside my home.

It might seem a bit redundant to use all three backup techniques, but that's the whole point.

#

WEBINARS

A glimpse into the future of education

January 22, 2016

In the software world there's a saying: "Eat your own dog food." It means using what you sell to your customers yourselves. For example, if your company developed a new web browser, then everyone in the company would use that web browser and not a competitor's product.

I've recently been dining on some puppy chow of my own.

One of the newer ways of delivering courses is through a webinar. The word "webinar" is a combination of "seminar" and "web," which implies a course that is delivered online, with the instructor and the students interacting through their web browser. It's live in the sense that everyone connects to the online course at the same time, irrespective of where they are geographically located. For an unlucky few, this means the course takes place in the middle of the night. But taking a course in your pajamas may be preferable to flying halfway around the world to attend the course in person.

Delivering a short webinar, say an hour or so, is not too difficult. But delivering a full course, such as a three-day corporate training session for six to eight hours per day, is much more challenging. The instructor is usually the only person speaking, and they are talking into empty space. They (usually) don't see the students; their only interaction is through short text messages posted to chat rooms. It's possible for students to speak, but that can quickly devolve into a cacophony of voices that are impossible to follow.

Cisco WebEx is a market leader for webinar delivery. It supports shared screens (where you see everything that's on the instructor's computer display), large classes, and limited student interaction. But I've yet to experience a webinar that doesn't begin with a tedious collection of technical problems: unsupported browsers, plugins not installed properly, audio functions difficult to setup (e.g., muting the student), and so on. This means webinars may be suitable for technically adept participants, but they may not be suitable for a more general audience.

Sitting in an advanced course for several days has been quite enlightening. At first, I found the instructor's delivery to be rather monotonous, but after listening to him for a while, and watching everything he has to do to keep the students engaged, I have a lot of sympathy for him. Webinars can also be recorded for later playback, but that eliminates any interactions between instructor and student.

Webinars are part of the future of education. With all their shortcomings, webinars are often the only way people can attend courses that fit their busy schedules. It's something we need to get right.

# # #

ED YOURDON AND MARVIN MINSKY

We've lost two computing pioneers in one month

January 29, 2016

This month the computing world lost two pioneers. Ed Yourdon passed away on January 20, and Marvin Minsky passed away on January 24. Both men made significant impact on the daily lives of millions of people – even if they didn't know it.

Ed Yourdon worked in software engineering. He introduced the world to structured analysis & design, a technique for developing software applications that is still taught in schools today. This contribution to the field led to his induction into the Computer Hall of Fame.

He was also a prolific writer who authored numerous textbooks. When I was in graduate school, he published the book "Decline and Fall of the American Programmer," which greatly influenced my own career choices towards academic research. A few years later, he published "Rise and Resurrection of the American Programmer," which gave some hope towards the future of the development profession.

Yourdon had somewhat alarmist views about the so-called Y2K problem. He felt that society's digital infrastructure was in real danger of collapsing due to the millennium date rollover. Thankfully, no such disasters took place – perhaps in part due to his warnings and the proactive changes made to many legacy software systems before 2000.

One of the last books Yourdon wrote was "Death March," which discussed the challenges of software project management with accelerated schedules on so-called Internet time. A central theme was the negative impacts such projects had on team members. I believe this book was part of the reason the agile movement became so popular shortly after its publication.

Yourdon received his undergraduate degree from MIT in 1965. Interestingly, MIT was also the home for Marvin Minsky for many years. Minsky was a professor emeritus at the MIT Media Lab and co-founder of CSAIL (the computer science and artificial intelligence lab). He is considered to be the father of artificial intelligence – a topic that has been much in the news lately. He built the first neural network simulating the human brain at Princeton University in 1951. In 1969 he received computing's highest honor: the Turing Award.

His books on AI and society became required reading for generations of students. "The Society of Mind" has been widely cited. I particularly liked one of this later books, "The Emotion Machine: Commonsense Thinking, Artificial Intelligence, and the Future of the Human Mind."

Minsky mentored many notable graduate students, including Ray Kurzweil, himself a noted futurist and author who has written about the possible outcomes of increased computing power to create an artificial intelligence with more capabilities than a human – the coming singularity.

Both Yourdon and Minsky will be sorely missed.

# # #

FACEBOOK TURNS TWELVE

Social media adolescence

February 5, 2016

On February 4, Facebook turned twelve. As a precocious adolescent, it sure has made its mark on society.

The company is worth more than $300 billion. Mark Zuckerberg, co-founder and current Chairman and CEO of Facebook, has become one of the wealthiest men in the country: his personal net worth is estimated to be over $48 billion. He's just 31 years old.

About 80% of Facebook's revenue comes from mobile advertising. Every time someone pulls out their smartphone to check their status, Facebook's cash registers ring. All you have to do is look around and observe people walking heads-down while gazing at their phones everywhere they go and you'll understand why Facebook makes so much money.

Facebook has nearly 1.6 billion users worldwide. That's almost one out of every four humans on Earth. Each day, a billion people use Facebook. Just sit back and think about that for a moment. The 2015 Super Bowl had an estimated television audience of 114 million – about 1/10 the number of people who used Facebook that day. Only the 2014 World Cup soccer final came close – and that was just for one day. Facebook is every day. And every night. And everywhere.

Can you remember a time when Facebook didn't exist?

Facebook wasn't the first social media website to the party. In

2003, MySpace was launched and that's where everyone wanted to be. But it didn't last long.

Facebook began as a tool to connect dorm buddies at Harvard. It soon expanded to create online social groups, giving a whole new meaning to being a "friend." Today, it's not uncommon for people to have thousands of online so-called friends.

A recent article in the Telegraph newspaper reported that there are 3.57 degrees of separation – not six. (I checked Facebook's research page and found my average degree of separation from everyone is 3.52.) The increased connectedness of modern society is due in no small part to Facebook.

For many people, Facebook has become synonymous with the Internet and the web. Rather like AOL used to be for people on dialup modems: a closed environment with little need to go anywhere else. Indeed, Facebook has become one of the largest sources of daily news for much of the population – particularly younger users.

It's interesting to speculate what might dislodge Facebook from its current position of prominence. New apps are released daily from hundreds of hungry startups, and consumers are notoriously fickle. Where's MySpace now?

In a dozen short years, Facebook has upended the world. It makes you wonder what it will do when it becomes a rebellious teenager.

# # #

VALENTINE'S DAY

Nothing says "I love you" like the gift of exercise technology

February 12, 2016

Nothing says "I love you" like the gift of exercise technology.

That's a good thing, because telling your Valentine that they are fat and/or lazy and need more exercise is probably not the best message to send on February 14.

"Wait a minute," you might say. "That's not the message I wanted to send. I just wanted to express how much I loved and cared about them, and how I want them to stay healthy and strong to live long lives. What's wrong with that?"

There's nothing wrong with that. Your intentions were good. Who can argue with a desire for your loved one to enjoy all the benefits that come from a healthy lifestyle?

Well, your Valentine, for one.

CNET reported that the number one mistake in their list of poor technology gifts for Valentine's Day is choosing exercise technology. It's just poor timing. Your message of health through technology is garbled during transmission, becoming more of an implied criticism of your Valentine's attractiveness. And Valentine's Day is supposed to be all about romantic attraction – not subliminal commentary on your partner's washboard abs that appear to be hidden by fluffy pillows.

Consider the gift of a digital scale. You might think it to be a vast improvement over the old mechanical one hidden in the bathroom. It's easier to read. It stores the results for later recall and comparison. You hope that its increased use will encourage your Valentine to watch their waistline. But few people like being given bad news, and if there's a need for your Valentine to lose weight, the first few weigh-ins are sure to report bad news (maybe even speaking the results out loud, for all to hear). Poor body image doesn't lend itself to romantic liaisons.

Another popular exercise technology is the Fitbit. It helps track your Valentine's steps every day. It provides helpful "Reminders to Move" when it detects a sedentary lifestyle. One might call such help digital nagging. When these results are uploaded to an online account, they provide an opportunity to analyze your Valentine's daily activities. You might like to think carefully before starting that discussion though. Not everyone likes having their movements tracked and critiqued. And it's hard to see how being told you're a sloth – by a device and by your partner – will make Valentine's Day more enjoyable for anyone.

Exercise technology has a place in modern life. But it has a time as well, and I don't think Valentine's Day is that time. Go old school instead: flowers, chocolates, a fancy dinner out. Save getting healthy for the next day.

# # #

APPLE, THE iPHONE, AND THE FBI

Personal privacy versus national security

February 19, 2016

Should Apple be forced by the FBI to unlock an iPhone?

A recent posting on the company's website by CEO Tim Cook called for a public discussion of this very important matter. So let's look at some of the arguments for and against Apple's compliance with the federal judge's order.

YES, Apple should comply. The iPhone in question was used during a terrorist attack in San Bernardino. The iPhone was owned by the county, not an individual, so an argument could be made there's less of an expectation of privacy.

Apple would also be seen as a "good corporate citizen" in the eyes of a nation shaken by world events that are often orchestrated using smartphones. There could be a feeling of relief among the general population that, while your personal data is safeguarded from hackers, legitimate government representatives could gain access to the data when they have probable cause and a legal search warrant to do so.

Apple could use the court order as a public relations shield, saying they were forced to comply by the government. They would be on the record as having tried to oppose compliance. But I suspect this case would go all the way to the Supreme Court before that decision was reached.

NO, Apple should not comply. Like many technology

15

companies, Apple has been quite vocal in their support of data protection and customer privacy. They claim even they cannot see your encrypted data stored on your iPhone. This has given them a certain level of appreciation in the market by people who feel strongly about such things.

If Apple does comply, there's no going back. Consumers will never again really trust any company who says their private data is secure. They will suspect a backdoor is installed on all of their devices – even when told otherwise. The technical details of how the iPhone is cracked are irrelevant to this discussion; only the results and public perception matter.

Lastly, how would Apple resist similar orders from countries like China?

SUMMARY. From a technical point of view, it probably doesn't matter either way what Apple decides to do. Reverse engineering an operating system, hacking some parts of the code to remove the data protection feature, and installing the modified program on the iPhone should not be beyond the capabilities of federal agencies like the NSA.

From a political point of view, if the government cracks the phone now, without Apple's explicit help, everyone will know the government can access their data. It's far better to follow the legal path for such a high-profile case. For now.

# # #

ENGINEERS WEEK

Three essential skills that all engineers need to master

February 26, 2016

It's National Engineers Week (EWeek) this week. According to the website of the National Society of Professional Engineers, EWeek began in 1951 and is "dedicated to ensuring a diverse and well-educated future engineering workforce by increasing understanding of and interest in engineering and technology careers."

Some engineering disciplines remain grounded in core principles that are ageless. For example, civil engineers still need to know some of the things about building bridges that the ancient Romans knew. But all of engineering is now suffused with computing, which means the rapid pace of technological change affects all engineers.

Universities constantly struggle to provide the right educational balance of a solid theoretical foundation with practical real-world knowledge. This is particularly true in the applied fields of engineering and computing. For this reason, I think there are three skills that all aspiring engineers should acquire while in school.

The first skill is the ability to learn rapidly. Learning how to learn is one of the most important things all students can benefit from in higher education. Consider a computer science student learning how to program. Whatever language they use in school, there's a good chance it will be replaced by a new language a few years after they graduate. So in addition to mastering the programming language in class, they need to learn how to learn a new programming language while on the job. They should be able to pick up a new language literally over the weekend.

The second skill to master is to become comfortable working with incomplete and sometimes inconsistent information. This is not easy for students in the hard sciences or engineering, because their work is grounded in facts and objective data. But the real world is much more nuanced. Conflicting priorities, different personalities from key stakeholders, and shifting requirements all tend to make project decisions very fluid. This is one of the reasons agile methods enjoy such popularity: they are tailored towards projects that acknowledge the inability to have all the facts before work begins. This is a departure from traditional methods, but I think it's essential today.

The third skill is a timeless one, and something that almost all advisory boards call out as lacking in recent graduates: the ability to communicate effectively. Oral and verbal communication skills are essential to a successful engineering career, but they are not areas of emphasis in a typical undergraduate program. I also believe that communication skills are even more important for senior engineers, because they serve as ambassadors to the general public. Learning how to explain complex topics to non-experts is not easy – but it's necessary.

# # #

PENTAGON ADVISORY BOARD

What can the government can learn from industry?

March 4, 2016

The Secretary of Defense recently announced the creation of an advisory board to the Pentagon, to "bridge the gap" between how the government and the technology industry operate. Google's former CEO, Erich Schmidt, will lead the board.

While I wait for my invitation to serve on the board to arrive, I thought I'd prepare my advice for Dr. Schmidt. I'm offering this advice for free, so remember: you get what you pay for.

Computing Literacy. My first suggestion is to ensure all federal employees have a modest level of computing literacy. This doesn't mean they all have to become top-notch coders. But it does mean they should possess a working knowledge of computing and how modern software systems work.

I find it ludicrous that some people seem to take a perverse sense of pride in their lack of information technology expertise. I doubt most office workers would say, "Call the English Department, I can't write this memo without their help," but some folks seem to have no problem with computer issues: "Call IT. I have no idea how this works. That's their job." True, the IT specialists have a role to play, but basic computing literacy should be a requirement, just as traditional literacy and numeracy are (hopefully) required.

Cybersecurity. Computer systems are incredibly complex, particularly those that rely on aging legacy systems. This complexity leads to vulnerabilities. It's often quite difficult to fully understand what's going on with interconnected programs, and this lack of visibility makes it easier for hackers to hide in the technical shadows. But difficult doesn't mean impossible.

Building upon a basic level of computing literacy, everyone should learn the basics of cybersecurity. There's no chance of making a system completely secure, but that doesn't mean we shouldn't adopt known best practices to avoid common pitfalls. Just as there's no way to completely secure your home, that doesn't mean you'd leave the front door unlocked when you go to bed.

Personal Accountability. A basic understanding of cyber security leads to my final recommendation of accepting personal responsibility for breaches and other forms of system failure. Security is everyone's job, not some faceless organization.

One of the stated goals of the advisory board is to "bring Silicon Valley innovation and best practices" to the government. One thing all entrepreneurs must accept is personal risk – it's the counterbalance to personal reward. You hear about startups becoming wildly successful all the time. What you don't hear are all the sad stories of the many startups that fail. But that's how innovation works. The trick will be getting the D.C. culture to adopt it.

# # #

COMMUNICATION MILESTONES

The telephone, hypertext and the web, and email

March 11, 2016

There were three big communication milestones this week. In the space of six days we celebrate the first use of the telephone, the birth of the inventor of hypertext, and the death of the creator of email. All three events are notable for their enduring impact on modern society.

On March 10, 1876, Alexander Graham Bell made the world's first phone call. The first words ever spoken were to his assistant, Thomas Watson, who sat in an adjacent room: "Mr. Watson, come here, I want to see you." Drawings of Bell's telephone apparatus show a rather crude mouthpiece and hearing device, but it worked. It was able to convert a person's voice to an electrical signal, transmit the signal along a wire, and convert the signal back to speech again. For something invented 140 years ago, it's hard to imagine another technology that has permeated life so thoroughly – with the possible exception of electricity.

On March 11, 1890, Vannevar Bush was born. Bush is credited with inventing what we now call hypertext. He worked for the federal government, advising the president on scientific matters. He published his famous paper, "As We May Think," in the Atlantic magazine in 1945. The paper outlined his ideas for the memex, a device to make knowledge more accessible by linking it by machine. These ideas formed the foundation for hypertext twenty years later, and the World Wide Web thirty years after that. These days, it's hard to imagine a time without the web.

On March 5, 2016, Ray Tomlinson, the creator of email, passed away at the age of 74. Tomlinson worked for Raytheon – a company that Vannevar Bush founded in the 1920s. In 1971, Tomlinson sent the first email on a computer network, introducing the now ubiquitous '@' sign to address the message. He said email was created to support asynchronous communication: you sent a message, and the recipient read it and replied later, rather like postal mail. He said the telephone was fine, but (lacking voicemail) it only supported synchronous communication: both parties had to be online at the same time. As for the contents of the first email, unlike Bell's famous words, Tomlinson says he doesn't remember what he wrote.

We now have more technically advanced means of communication. It's commonplace to use streaming video to hold meetings with colleagues who are separated by great distance but who are brought together by a virtual environment that lends the illusion of everyone being in the same room at the same time. However, even these modern applications rely on telephony, the web, and email addresses to work properly.

# # #

EFFECTIVE USE OF WRITING TECHNOLOGY

People + programs create better stories

March 18, 2016

At the Brevard Authors Book Fair this weekend I'm giving a talk called "7 Habits of Highly Effective Writers." Notice the lack of the definite article in the talk title. I'm not presuming that there are only seven habits, or that I've identified the canonical set. But through my own personal experience I have come to recognize certain character traits that I believe can make an author more successful, and one of them is effective use of technology.

Writing today still centers around the essential act of putting pen to paper – except it's not always a pen, or paper. You still need to learn and practice the craft constantly, which is why courses on core writing activities will always be needed. But as with all modern activities, technology has changed how these activities are performed.

I'm probably one of the few computing people who still write longhand in a daily journal. My calendar is a Brownline book, not an app. But the vast majority of my professional writing is done using Microsoft Word, and since this is the main tool I use, I've taken the time to learn how to use it properly.

Word is a complex program that annoys me almost daily – but I still have to use it. Most people barely scratch the surface of Word's features. For writers, Word offers many timesaving capabilities; the problem is that it's not easy to master them. For example, I'm constantly surprised when someone submits a manuscript for me to edit as part of a book I'm working on, and I see how they have completely ignored (or misused) Word styles. This leads to increased labor when reformatting the content. You don't need to love Word, but you do need to live with it.

When it comes to editing, I still rely on human experts. I find their insights invaluable. A second set of eyes provides much-needed perspective, identifying awkward passages that I completely glossed over. Technology even helps here: there are many websites that list editors offering their services, complete with pricing and reviews.

I also make use of automated editing tools, such as Grammarly, to compliment the human editor. Sometimes the person misses a copy edit; sometimes the program makes the wrong change. As the writer, the final decision is always yours.

I've found that the software powering the editing programs to be quite sophisticated. I often have to do further research to determine the origins and validity of the suggested changes. Sometimes the grammatical rules are quite nuanced. But there's no doubt that I've become a better writer by using these tools.

#

TWITTER TURNS TEN

Follow me @TechTodayColumn

March 25, 2016

On March 21, Twitter turned ten. With about 300 million users, Twitter has never become as popular as Facebook, but it certainly has had an impact.

Jack Dorsey, co-founder and current CEO of Twitter, sent the very first tweet in 2006: "just setting up my twttr." It doesn't have the same gravitas as Bell's first telephone message, but it did set the tone for the billions of 140-character messages that would follow in the next decade.

As of today, I've sent out 292 tweets through my @TechTodayColumn username. "Please follow me!" is the usual refrain to get people to subscribe to your tweets. With this account, my tweets are pretty limited to announcing when a new column is published. Not the most exciting, I know.

I use the hashtag #TT to index the Technology Today columns. Hashtags are simple keywords that computers can use to search for specific tweets in the Twitterverse. Both the '@' and the '#' symbols were introduced to Twitter organically: someone at Twitter thought it was a neat idea, and it caught on.

Twitter is commonly used in times of emergency to send out brief messages, almost like online flares. But just like YouTube has become the main source of cat videos, Twitter has become a prime source of vapid commentary. Here's a recent tweet by presidential candidate Donald Trump (who has over 7 million followers): "@Don_Vito_08: 'A picture is worth a thousand words' @realDonaldTrump #LyingTed #NeverCruz @MELANIATRUMP." Content aside, I find it hard to parse these messages, full as they are of '@' and '#' references to other people and trends.

What I read in the news a lot are statements like this: "Then the rivals got into a Twitter exchange…" In other words, tweeting has become the digital version of mudslinging. That's hardly the most inspiring use of technology. But it seems Twitter users are not too interested in substance. According to TwitterCounter.com, the top three accounts (in terms of followers) all belong to entertainers: Katy Perry (84 million), Justin Bieber (77 million), and Taylor Swift (73 million). For comparison, President Obama comes fourth with 71 million. That tells you were our priorities are.

People can show their interest in a tweet by retweeting it – sending it to their followers. According to CNET, the most retweeted tweet was in 2014: Ellen DeGeneres' famous Oscar selfie. That is, until the 2016 Oscars, when the award for Best Actor went to Leonardo DiCaprio, which generated an astounding 440,000 tweets a minute.

Was there really nothing more important going on in the world on February 28, 2016?

# # #

APRIL FOOLS

Which is real and which is fake?

April 1, 2016

Technology today advances so rapidly that sometimes it's hard to know which new developments are real and which are not. To celebrate April Fools' day, guess which of the following three technology stories are real, and which are fake?

FBI Offers iPhone "Hack for Hire." The FBI has successfully hacked the iPhone used by the San Bernardino terrorist. This was a surprising development, since they had initially tried to use the courts to force Apple to circumvent the encryption. When Apple refused, the government seemed resigned to pushing the issue through the legal system. Then, this week the FBI said an "outside source" had approached them with an offer to crack the phone. Apparently it worked, because the FBI withdrew the court order and said they had access to the phone's data.

Now comes news that law enforcement agencies across the country as asking for the FBI's help in cracking phones they have in their possession. For example, the FBI has agreed to unlock an iPhone and an iPod belonging to teenage murder suspects in Arkansas. Presumably they will do this using the same technology they used for the California case. This won't make Apple very happy, since they have publically stated that their system was impenetrable.

Is the FBI now offering a "hack for hire" service?

Scientists Create Insect-Computer Hybrid Robot. Reports

from Singapore suggest scientists have managed to create an insect-computer hybrid robot. This is a real, live insect (a large beetle) that has its nervous system connected by tiny wires to an external microcontroller. The controller is able to directly manipulate the bug's limbs, causing it to walk where it is directed.

This sort of blended organism could have many uses. For example, insects are able to crawl through very tiny spaces. They could have sensors and even a small camera mounted on their body, which would let first responders assess emergency situations very quickly. Forget about drones; cyber-dragonflies are the future.

But is turning beetles into Borg really a positive development?

Japanese AI Program Writes Novel, Almost Wins Literary Award. A Japanese computer program that used artificial intelligence to "write" stories almost won the Nikkei Hoshi Shinichi Literary Award. The story was judged in a blind review process, which means the reviewers did not know if the author was man or machine. The lead researcher behind the program said the goal was for the AI system to resemble human creativity.

How long before the Nobel Prize for Literature is won by a robot?

* * *

All three stories are true, and that's no April Fools' joke.

\# \# \#

ARTIFICIAL EVOLUTION II

Robots making robots

April 8, 2016

In August 2014, I wrote a column about artificial evolution. The focus was on the creation of new forms of life, biological or digital, which would then evolve at incredible speeds. The emphasis then was on the software part of life: the algorithms that gave the appearance of intelligence.

But what about the hardware side of evolution? Can artificial life physically beget artificial life? Can it produce offspring?

Recent reports from MIT's CSAIL (computer science and artificial intelligence lab) indicate that researchers were able to create a working robot using a 3D printer. Creating inexpensive robots in itself is not terribly novel. A few years ago, a Harvard/MIT team was able to create paper robots that unfolded and started walking by themselves. They were rudimentary, but they were cheap and simplistically autonomous.

These new 3D printed robots are different. By "working" I mean the robot was ready to go as soon as it was printed. Just add a battery and a small motor and it starts to walk. No assembly or training required. These robots are rather like a young doe that wobbles on its legs shortly after birth, but is quickly able to stand and then walk without being taught how to do so. Except the robots skip the wobbling phase.

By "printed" I mean the robot was created in a single process on a Stratasys 3D printer. The robot is literally created from raw materials (both solid and liquid) in one pass. It moves using a hydraulic system of six legs. Again, everything is printed; nothing is added to the robot after it emerges from the printer (other than the battery and motor). This includes the liquids needed to power the printed crankshaft and the robot's insect-like legs.

It's not too difficult to imagine the newly created robot being able to hit the "print" button and create more robots. At that point, we've arrived at a form of robotic procreation. Artificial evolution is getting closer all the time.

As with all artificial processes, there's nothing to say each new robot design could not be an improvement on the previous design. All that would be needed is a bit of programming that seeks to continuously optimize certain robot characteristics, such as gait or strength. Once we can print solar cells or some other energy source, even the batteries would not be needed. Perhaps the motor could be printed too. We'd literally have mechanized life coming out of the womb of the future.

Now partner the previous work on AI evolution with the robotic birthing process, and we're close to creating our overlords. Is this wise?

#

CLIPPY

How may I help you?

April 15, 2016

Do you remember Clippy?

It's been 15 years since Microsoft discontinued its office automation assistant. Clippy was an animated character that looked like a giant paper clip with big eyes. He appeared without warning on your screen and offered to help with your current task.

For example, if you were typing a business memo, the Office software could track and analyze what you were doing, and if it detected simple patterns, Clippy would pop up. It could fill in some fields, based on system templates, which in theory would make the task of writing the memo easier.

In practice, Clippy's unwelcome appearances were soon the subject of much derision. People spent more time making Clippy disappear than they liked. There seemed to be no way to turn this "feature" off. In the end, Clippy was removed from the next version of Office – an interesting idea, poorly executed. Fast-forward to today, and the central concepts behind Clippy are being used again, but this time in a much more effective manner.

At the Big Data Florida user group meeting this week, Syam Suri of Verizon spoke about how the company uses real-time analytics to improve the customer experience. Just as Clippy would lurk in the Office background, watching what you did in Word, Verizon is now able to monitor customer usage patterns to deliver improved service across multiple channels. For example, if you are an existing customer on their website, looking to purchase a new smartphone, and then call them to place an order, they should be able to pick up the conversation from where you left off with the website, rather than starting from scratch.

This type of insight is made possible by software programs that process data that is flowing into Verizon's systems from their customers. It's like Clippy was trying to do, but at enterprise scale. But it's unlike Clippy's user interface: there is no visible change to the workflow for the customer, other than to expedite matters.

Verizon is not the only company doing this of course. Many large companies are trying to harness the power of big data to improve their operations. The business drivers to do this are obvious, but the technology to make it a reality is non-trivial.

There is always an issue of privacy when it comes to customer data and analytics. Many people found Clippy a bit creepy: the way he was watching what you were doing was off-putting. At the same time, many consumers seem willing to let their personal data be used – if it's done properly. If this were not the case, we wouldn't have Facebook.

# # #

EARTH DAY

Apple's iPhone abattoir

April 22, 2016

One of the banes of modern electronics is that so much waste is produced when we upgrade to the latest version of our gadgets. The Electronics TakeBack Coalition reports that 20 to 50 million metric tons of e-waste are generated worldwide every year. This is terribly inefficient, but it's worse than that: a lot of the waste is toxic. Mashable reported that electronics are responsible for over 70 percent of the toxic waste produced in the U.S. each year. The toxicity comes from sources like rare earth elements used in the circuitry and batteries of the devices.

Many companies have tried to recycle their electronic components, but it's devilishly difficult to do properly. Supply chains span the globe. Not everyone wants to go through the hassle of returning their devices to the store – and we're a long way from a general "Blue Bin" approach. The very act of properly disassembling gadgets like smartphones is complex and laborious; after all, these aren't newspapers and tin cans. Thankfully, we have a workable solution for laborious and repetitive tasks: robots.

Liam is the name of Apple's iPhone 6S robot recycler. This 29-arm autonomous robot basically runs Apple's digital abattoirs in California and The Netherlands. It can disassemble an iPhone in just 11 seconds. That sounds fast (and it is), but it also means that Liam can process just a small fraction of the iPhones that Apple sells. In 2015 alone, they sold over 230 million, and there are over 1 billion Apple devices currently in use worldwide. But Liam is a promising start. You can read more about Apple's environmental efforts online at http://www.apple.com/environment/.

Ultimately, designing and manufacturing electronics that are inherently more reusable would be beneficial, both environmentally and financially. But that will require techniques that don't rely on toxic components; they are used due to their unique capabilities, not because they are inexpensive. I think that as with most advancements in sustainability, long-term success will only occur when it makes good economic sense.

Apple is an exemplary corporate citizen. They use renewable energy in their US facilities. They replant trees to combat emissions. Their new headquarters is said to be a model leading-edge eco-friendly set of buildings when they open in 2017.

But Apple is also making a nice profit with Liam. Apple reported that during fiscal 2015, Liam was able to recycle over 2,000 pounds of gold from iPhones. That's a tidy little pot of nearly $40 million of gold that otherwise would be languishing in a landfill somewhere.

Liam has a very particular set of skills indeed.

# # #

AFTERLIFE AVATAR

Soon you may never have to let them go

April 29, 2016

I've heard about people keeping stuffed versions of their deceased pets in their house. It gives them comfort to see a familiar figure sitting in their favorite place. Personally, I find having a freeze-dried version of Fluffy a bit extreme, but who am I to judge.

I've also heard about people who clone their pets. They use a process of genetic preservation to duplicate the DNA of their current pet, to create a copy of the dog or cat. The assumption is that many of the traits of the original pet will be carried over with the cloned pet, although that does raise all sorts of "nature versus nurture" questions.

Pets are one thing, but what happens when we start preserving the past in similar ways with humans? Would you do it? If someone told you your loved ones could be brought back into your life, could you say no?

Technically, I don't think we're there yet. Ethically, we may never be. But imagine we could do so virtually. Imagine we could create a realistic avatar of someone. They would "live" online, even after their death, but you could communicate with them. Would you do it?

This question is the essence of "Be Right Back," an episode of the British television series "Black Mirror" that I recently watched. It was incredibly thought provoking. More importantly, it didn't seem too far-fetched; our technology is almost there and improving rapidly.

In the show, a young lady loses her boyfriend unexpectedly. She learns about a new online service that will let her communicate with him via email and instant messages. At first, she's reluctant to try it, but when she eventually succumbs to the temptation, she finds the experience quite realistic. The software that powers the digital version of her boyfriend has scoured the Internet for copies of his public emails and messages, analyzed the text, and created a model of his online persona. The program then says it can be improved if it has access to her boyfriend's private photos and videos. Quite quickly, the avatar can interact with the woman in ways that seem shockingly real.

Can we do this today? No, but artificial intelligence algorithms, big data analytics, and powerful computers can create sophisticated models of real-life scenarios already. What's to stop someone from creating an app that mines the past lives of someone who spent a lot of time online, and creating a simulacrum of them?

At that point, it's only a small step to embedding this programmed behavior into a human-form robot.

When that happens, I wonder who will have the strength to say no?

###

WHAT I LEARNED IN EDUCATION

The irresistible force meets the immovable object

May 6, 2016

This week marks the end of my formal stay in a university education department. In the Fall 2016 semester I'll be returning to the College of Engineering. As is the case whenever some significant change occurs, it's instructive to summarize some of the key lessons learned from the experience.

What I learned from my tremendously enjoyable time working in education is that making changes in the field is an extremely challenging task. This is hardly a headline; the news is full of stories of the difficulties in effecting reforms. But for me, it became personal: if technology is the irresistible force, education is the immovable object.

The first thing I learned was related to the struggles of teaching pre-service STEM educators about the role of computing in the future of education. There are two traditional ways science advances: theory and experiment. Computing has added a third path, and this new method carries over into the classroom. Computational science now permeates subjects like biology, mathematics, and physics. This implies educators must become sufficiently proficient in computing to use these new tools with their students. As much as I tried, I just couldn't seem to get these teachers-to-be to understand that computing was going to transform much of STEM education. But this is not something I will ever give up on; it's too important.

The second thing I learned is that there's a conundrum related to computing education at the K-12 level. One the one hand, everyone from the federal government on down recognizes the pressing need for more computing educators. Indeed, the National Science Foundation has several programs, such as the CS10K initiative to create 10,000 new teachers across the country, specifically designed to address this strategic shortage. On the other hand, the educational system seems setup so that actually getting in-service teachers to become computing literate is virtually impossible. Their schedules are already full; they lack proper incentives to gain the training and certification necessary; and the piecemeal approach adopted by various school districts and state agencies makes rolling out new programs at scale problematic. But the need for computing educators will only grow over time, so it's in our national interest to address it.

The third thing I learned is that teachers and education researchers are truly passionate about their work. They know that what they do directly affects people's lives and they take that responsibility very seriously. We just need to give them the tools to succeed. Sometimes that's new technologies to be used in the classroom. But sometimes that's just the freedom to do what they do best: focus on teaching their students

# # #

RANSOMWARE

Who you gonna call?

May 13, 2016

My email inbox has been flooded with more spam than usual lately. The messages look official. They usually address me by name. The subject line references an unpaid bill, a list of vendors, pending deposits, and other topics that sound legitimate. The message always includes an attachment as a zip file. I delete the messages, but it's tiresome because it must be done manually each time.

If you've received similar messages recently, I hope you delete them too. They aren't merely annoying. They are in fact phishing attacks that can be launched when you click on the attachment and open it. Once that's done, it's too late: you're infected.

A phishing attack is a simple trick to install malware (e.g., computer virus) on your computer. When you click on the attachment, the malware installs itself. The malware payload can be almost anything, from keystroke loggers to fake websites. Anti-virus software can help, but the best defense is being aware of the prevalence of phishing attacks in the first place.

There's been a new surge in a particularly nasty type of malware called ransomware. As its name suggests, ransomware will literally hold your computer hostage until you agree to pay a ransom. The program encrypts your data, such as all your documents, so that you can no longer access them. You are forced to pay a ransom to get the key to unlock your own files.

As reported by Wired a few months ago, the Hollywood Presbyterian Medical Center in Los Angeles was hacked using a ransomware program called Locky. The hospital's computers were unusable for several weeks. Officials eventually paid the $17,000 in Bitcoins demanded by the attackers. Bitcoins are a preferred method of payment because the digital currency is all but untraceable.

This week, TechCrunch reported that the House of Representatives are being targeted with ransomware using phishing attacks though YahooMail. Legislators and staffers have been warned against opening attachments or clicking on links embedded in email if they don't recognize the sender. And even then, the sender's identify can easily be spoofed.

It's a pity that smart minds are wasted creating inventive programs to hold innocent user's data hostage. But the online world is no different than the real world in this respect. All you can do is try to protect yourself – in both places. Practice good cyber hygiene, be cognizant of phishing attacks, and above all, have backups of all your data in place. It may turn out that doing a complete system wipe and then re-loading your data files is the only alternative to paying a ransom.

# # #

THE LONG SAD SUMMER

Challenges in revising technical textbooks for today's students

May 20, 2016

This weekend I'm giving a talk called "The Long SAD Summer." The talk title may suggest that my summers are unhappy times, but that's far from the truth. The acronym "SAD" refers to "Systems Analysis and Design," the code I used when writing the 11th edition of the textbook by the same name for Cengage Learning over the past two years. The "Long Summer" phrase is an allusion to an old song by the band "The Style Council" called "Long Hot Summer." During the summer I write in Arizona and Florida, so the temperature reference seemed appropriate.

Systems analysis and design is a complex topic. It includes a lot of technical material that students must master in a single semester. It also includes a significant amount of non-technical content related to topics such as project management, return on investment analysis, and communication skills.

As with most computing-related textbooks, SAD had grown over the years. More material was added to each new edition to reflect the rapid changes in the field. The result was that the 10th edition of the book had ballooned to over 650 pages. I've taught a class based on this earlier edition, and it's a veritable fire hose of information for the students to absorb.

As the author of the 11th edition, one of my primary goals was to streamline the book's content to make it more approachable to today's students. I still included expanded coverage of emerging technologies, such as agile methods, cloud computing, and mobile applications to compliment traditional approaches to systems analysis and design. I also used a lot of real-world examples to emphasize critical thinking and IT skills in a dynamic, business-related environment.

To balance the new material, I made a decision to judiciously remove a lot of older material. Some of the older material was simply deleted; other portions were moved online. The result was a revised edition that is 100 pages shorter than the previous edition.

Textbooks today also have a significant online component; the book itself is just part of the overall educational platform. This is done in part to support the instructor, but also to reflect the preferred habits of today's learners. Students are used to getting answers online. They want bite-sized pieces of information with little distraction. They are very task-oriented learners, particularly in business and engineering.

The book has 12 chapters and 4 appendices (called "Toolkits"), which makes it appropriate for a typical 16-week semester. But covering all the material still asks a lot of the instructor and the students. I will continue to address this challenge in future editions of SAD.

# # #

SUMMER READING

Fun, provocative, and philosophical

May 27, 2016

For my suggested summer reading list this year, I've selected three books that offer insight into very different subject areas. Depending on your mood, you can enjoy a book about a long-lived franchise in the movie industry, you can consider the ethical implications of moving away from fossil fuels, or you can ponder universal values and the state of education today. It might not seem so, but technology plays a key role in all three books. So take your pick (entertainment, environment, or enlightenment) and head to the beach.

The first book is *Some Kind of Hero: The Remarkable Story of the James Bond Films* by Matthew Field and Ajay Chowdhury (The History Press, 2015). This is a 700-page tome that provides fascinating insight into the James Bond franchise. Beginning with Dr. No in 1962 and culminating with SPECTRE in 2015, the book covers each of the Bond movies in great detail. I particularly enjoyed reading about the writing process for each movie, and the various specialized roles played by key personnel. I learned a lot about how large productions are managed and the role of technology in making some of the incredible special effects a reality on screen.

The second book is *The Moral Case for Fossil Fuels* by Alex Epstein (Portfolio, 2014). The current discussion concerning climate change and the role fossil fuels play in a modern economy is heated, to say the least. This book encourages the reader to consider the ethical implications of moving from fossil fuels to sustainable energy, especially given the limitations of current technology. I found the book to be very thought provoking. This is a complex topic that demands one take a more open attitude to the scientific issues and moral dimensions of providing an acceptable quality of life for billions of people.

The third book is *The Abolition of Man* by C.S. Lewis (HarperOne, 1944). This is the most philosophical of the three books on my list, and the oldest by far, but it's also one of the most prescient treatises on the long-term effects of education on society. Lewis was a university professor, so his views on public education are grounded in his own experience and observations. The book is a challenging read, but Lewis' commentary on universal values is strikingly relevant today. The news is full of stories about unapologetic cultural diversity and fluid moral relativism, driven in part by mass migration and political correctness. Rapid technical advances, particularly in medicine, will make the conversation about scientific progress and what we collectively consider to be "good" and "bad" cultural impacts even murkier.

# # #

iPhone in the Pool

That sinking feeling

June 3, 2016

I knew it would happen eventually; it was only a matter of time. Last week my iPhone went for a swim. One minute it was sitting on a towel beside me and the next minute I was watching it sink to the bottom of the pool.

Fortunately, it fell in the shallow end and I was able to retrieve it almost immediately. Unfortunately, "almost" still allows enough time for the phone to be well and truly submerged. It's not like the water needs a few minutes to tunnel into the device; it pours into the various buttons and connectors immediately and voila! Circuitry meets salt water – with predictable results.

The first thing I did was dry off the phone with the towel, and that seemed to work. Only after I finished drying it did I realize that the phone was actually still on. A tiny flicker of hope – maybe the phone would be fine? I tried using it and it worked. Success!

I turned my attention to the reason I knocked the phone into the pool in the first place: my cats. They had been chasing one of Florida's many lizards and caught one of the poor things. I had intended to rescue the lizard from the cat's mouth and place the lizard outside the enclosure. (This is a daily ritual.) But when I got out of the pool to chase the cat, I grabbed the towel, forgetting that the phone was hidden in the folds to keep it out of the hot sun. In retrospect, a little tan would have been better for it than a little swim.

Confident with the knowledge that the phone was fine, I put the whole incident out of my mind. Until the phone stopped working a few hours later and my flicker of hope was doused.

I quickly did what everyone else does when expensive electronics stop working: I yelled at it. The phone didn't respond. None of the buttons were working, which indicated that water had indeed gotten into the phone. I then remembered there's a high-tech solution to the problem: rice.

So like thousands of people before me, I placed my iPhone in a plastic baggie and filled the bag with uncooked rice. I then left the phone sitting on the counter for nearly a full day. It's amazing that a commodity like rice and a cheap plastic bag can rescue an $800 phone, but it worked. The rice draws out and absorbs the moisture very nicely. When I took the phone out of the bag, it was working again. Hurrah for low-tech solutions to modern problems.

# # #

1956

60 years of changes

June 10, 2016

This week my parents celebrated their 60th wedding anniversary. It's a rare event these days for couples to make it to their diamond year together – so rare that Hallmark's list of anniversary gifts stops at 60. But my folks made it, and I was honored to throw a little party for them. I had organized their 25th and their 50th wedding anniversary parties as well, so it was wonderful to be able to continue the tradition.

When family and friends get together for such a special occasion, it's natural to reminisce about the past. Way back in 1956, when my parents were married, the music world was introduced to Elvis Presley when he made his first appearance on the Ed Sullivan show. On TV, the game show "The Price is Right" made it's debut. In transportation, gasoline cost $0.22 a gallon. Listening to everyone talk about these things was interesting, but my thoughts naturally turned to technology.

The differences in computing capability between 1956 and 2016 are vast. Processors today are orders of magnitude faster (and smaller). In 1956, the Internet was still decades away. User interfaces were crude, text-oriented, and still relied on punched cards. Access to computers was limited to programmers and operators. But to really appreciate the advances technology has made in 60 years, consider storage systems.

In 1956, IBM introduced the first commercial computer with a hard disk drive (HDD) storage system. The 305 RAMAC computer had a HDD called the 350 that had a 5MB capacity. The 350 HDD alone weighed over a ton. Physically, it resembled two refrigerators encased in a metal shell. According to the EE Times, the RAMAC HDD used "50 hefty aluminum disks coated on both sides with a magnetic iron oxide, a variation of the paint primer used for the Golden Gate Bridge." Discs rotated at 1,200 rpm. The system was controlled with vacuum tubes.

In contrast, consider today's storage systems. Personal computer HDDs with 3TB of storage capacity can be purchased for about $100. The discs rotate at 7,200 rpm. The system is controlled by microelectronic circuitry. Enterprise storage systems that manage big data repositories with petabytes of information are quite commonplace.

Even more impressive are the solid-state storage devices I now use. There are no more spinning discs – just memory chips. They are smaller than your fingernail, yet they offer 512GB capacities with nearly instantaneous retrieval times.

There have been many significant changes in our society since 1956, led by the incredible innovations in technology. But some things haven't changed. My parents are still together. And "The Price is Right" is still going strong.

###

CONNECTED CARS

The radio richness of the mobile Internet of Things

June 17, 2016

Imagine you could put on a special pair of glasses that would let you see radio waves. For example, if you had a cell phone in your hand, the glasses would show the signals coming and going from the device, rather like digital fireflies zipping back and forth. Even for a cell phone you'd need different colored fireflies to truly visualize the radio traffic, because modern smartphones use many types of radio signals to communicate: cellular (GSM, CDMA, and so on), Wi-Fi, Bluetooth, GPS, and NFC to name just a few. All these signals would look like a multi-colored rainbow arcing out of your phone and disappearing into the ether.

Now imagine how much richer that radio rainbow would be for a more complicated device: a modern automobile. Today's connected cars are the epitome of the "Internet of Things" (IoT), except that they're also mobile. The complexity of modern car communications systems became very apparent to me recently while I was shopping for a new vehicle. I eventually settled on a Honda CR-V (Touring edition), but it's by no means unique in the nature of its radio signal richness.

Consider just the central navigation system, which you interact with via touchscreen controls. The system contains a radio, so there are signals for AM, FM, and the HD versions of both. There is also a Bluetooth link to connect to your smartphone. There is a satellite link for Sirius XM radio. There is a GPS receiver for the mapping software. Each of these different types of signals is deeply connected to the car.

Then there's the car's radar system, which is used to monitor distance to cars traveling ahead of you. If you get too close, based on current speed, the car warns you to slow down. There are also multiple sensors beaming out of the car to watch the lane indicators; if you stray from your lane, the car warns you to re-center the vehicle. Some of these sensors are cameras, such as the backup and turn warning systems, so they rely on the visual portion of the electromagnetic spectrum, but from the car's point of view they are just another signal to process by the control system in real-time.

If you were wearing your special glasses, the incredible number of radio signals surrounding and penetrating the CR-V car might blind you. But if you were driving the Quadricycle, Henry Ford's first car built 120 years ago, your special radio glasses would be dark. Maybe that wouldn't be so bad: you'd be more focused on driving. Especially since the Quadricycle didn't have any brakes.

# # #

WATSON

Canine versus artificial intelligence

June 24, 2016

I've finally had a chance to play with Watson. I've been reading about "deep learning" technology and how it can be used to help Watson master new skills with little outside help. Watson's neural networks can model problems in the real world quite rapidly and with high fidelity. There seems to be little doubt that past experience plays an important role in solving whatever problem is put to Watson: the first time it takes a while to react properly, but each time afterwards the result gets better and better. The more data is provided, the better Watson's understanding of the situation becomes.

Well, that and being given treats and told "Good Boy!" whenever he does something right.

Watson is my new dog. He's a 12-week old Golden Retriever.

Although he's just a puppy, Watson has already learned several important skills, such as how to "sit" when told. His brain has registered the pattern: sit and be rewarded with a tasty tidbit. He's learning more patterns every day.

Watson is named after Sherlock Holmes' sidekick, John Watson. The fictional Watson was a faithful companion to his more cerebral partner. He was also adept at providing Sherlock with new perspectives on crimes they solved together – a more intuitive approach based on Watson's understanding of the human character. Faithfulness and empathy are two positive characteristics that most dogs share. (I also used to have a cat called Moriarty, so it seemed appropriate to continue the pet association with Conan Doyle.)

Watson is also named after IBM's Watson supercomputer. Not that the dog shows much predilection for predictive analytics, although he has become quite good at navigating the house and darting to the door to go into the yard for his frequent walks. The computing connection for his name is based on the choice of names for my two cats, Hopper and Ripley, both of whom were named after famous strong female characters with a connection to computers and technology.

So how different is Tilley's Watson from IBM's Watson? Well, the former represents old-school canine intelligence (CI), while the latter represents new-fangled artificial intelligence (AI). They don't really compete with one another. CI is wetware (lots of slobbering), while AI is hardware and software (lots of programming). CI likes Milk Bones, while AI likes ... actually we're not sure what AI "likes" per se – it doesn't display any emotion at all.

IBM advertises their Watson as a solution to many problem domains, using phrases like "cognitive finance" to "outthink risk." My Watson is just a cognitive canine that likes to chew on things. The only risk is to my shoelaces.

#

BREXIT

How does the UK's departure from the EU affect technology?

July 1, 2016

The decision by the United Kingdom to leave the European Union has caused more ink to be spilled by pundits worldwide than almost any recent global event. Although the final outcome of the so-called Brexit probably won't be known for several years, that hasn't stopped analysts from chiming in on all aspects of this messy political divorce. Well, I don't want to feel left out, so I'll opine on the technology implications of Britain's landmark decision.

Most of the mainstream commentary on Brexit has focused on perceived negatives for the UK. For British technology companies, particularly London-based financial technology (fintech) companies serving The City, the mood appears to be particularly grim. There seems to be a consensus that many of these fintech companies will no longer be able to compete in the European market, due in part to a return to restrictions on movement of personnel between EU countries.

To use an appropriately ribald English saying, what a load of cobblers.

The intellectual capital that infuses London is not going to evaporate overnight just because the UK is pulling out of the EU's political project. Economic ties between Britain and the continent will remain – it's in the best interest of all parties to make it work. Indeed, supporters of the Leave campaign argue that Brexit will open new opportunities for fintech companies to expand to non-EU locations that have much higher growth rates than the sclerotic old country, such as India and China.

Consider another world financial center closer to home: New York City. Wall Street seems to get along just fine without being part of the EU. There are problems with work permits, but talent inevitably finds its way to the big investment houses in the Big Apple. And there's no shortage of startups in New York either.

Part of the reason startups thrive alongside established businesses is that new business models have disrupted how work gets done: there's no longer a fundamental need for everyone on the team to live in the same place. Most startups fully embrace the idea of geographically distributed employees. For example, there may be a few people in London, a few in Warsaw, a few in Chennai, and a few in Sao Paulo. Brexit is not going to change anything for these forward-looking companies.

Above all, money talks. Why would EU partners not want to conduct commerce with the 2nd largest economy in the bloc (and the 5th largest in the world)? After Brexit, there may be a few more administrative details to sort out, but UK fintech companies may be perfectly positioned to help make that happen.

#

NAPSTER

What has changed in 15 years?

July 8, 2016

Back in the day I had a pretty significant record collection. But that didn't stop me from making tapes of my friends' records; I always wanted more music. I remember the challenge of recording each side of the vinyl onto TDK 90-minute cassettes, hoping that the songs would not get cut off if the tape ran out. Long Pink Floyd cuts were always problematic.

When CDs came out I mostly stopped buying records. The introduction of CD-R recordable discs and the MP3 music format put an end to my cassette collection too. But I continued to purchase music on a physical medium. Digital music made piracy a lot easier, but real discs were still sold on the black market; the distribution was physical.

When Napster came out in 1999, everything changed. It was a prime example of disruptive technology. Distribution moved online, permanently altering the digital music landscape. All of a sudden, people could download just about any track or album they wanted – for free. The music industry never really recovered.

Napster's technical flaw was that it stored pointers to the music files held on other people's computers on a central server. If the server went down, so did the service – which is how the courts were able to shut down Napster in 2001.

Napster's legal flaws were obvious: it enabled music piracy on a scale never seen before. Musicians were understandably upset that people around the world were stealing their hard work. There were no contractual mechanisms in place for artists to be fairly compensated. Fans mocked popular bands like Metallica who were quite vocal in their opposition to Napster: a video cartoon of Metallica complaining "Napster, bad" was very popular at the time. Eventually the industry did embrace digital music, but it took Apple and iTunes to do it.

Napster would have disrupted the movie industry too, but the studios were saved primarily by one technical issue: bandwidth speeds were so slow back in 2001 that downloading the gigabytes necessary for a full video took too long. Music files are a lot smaller, so even dialup modems could manage it.

Napster has been gone for 15 years, but online media piracy has not gone away. The centralized storage model of Napster has been replaced by a decentralized system of torrents (downloadable links to files) that is much more difficult to shut down. The Pirate Bay, one of the main torrent websites, has proven to be more nimble than the courts.

People's attitudes have not really changed that much either. HBO's "Game of Thrones" is purported to be the most downloaded TV series in history.

# # #

INNOVATION IN THE REAL WORLD

The fear of failure is a very real concern for most people

July 15, 2016

Innovation is a perennial topic in the technology sphere. Without innovation, we would not have the exciting new products and services that so many amazing companies introduce to the marketplace.

There have probably been more books on innovation written than anyone could possibly read in a lifetime. The very fact that new books on innovation are constantly published indicates that we've not really solved the fundamental problems that make it so challenging. So what makes being innovative so difficult?

Innovation was the central theme at "One Nation: American Innovation" hosted by FLORIDA TODAY and the USA TODAY network and presented by Harris Corporation.

Mark Randall, Adobe's vice president of innovation, said, "The opposite of creativity is fear – the fear of failure."

I agree; the fear of failure is a great barrier to innovation. But I don't think it's the fear of trying a new idea and having it fail that's the problem. It's the fear of the reprisals and negative consequences most people suffer when something goes wrong that stops them from thinking and acting outside the box. Once burned, twice shy – and career burns can be deadly.

In a startup environment, there is a small group of like-minded people who are inherently predisposed to taking risks. They accept that most attempts at innovation will fail; it's the successes that are rare. When a failure occurs, the company pivots to a new idea or a new approach and just keeps soldiering on, confident that they will eventually crack the nut.

But in the real world, and in particular in areas like the Space Coast, most people don't work for startups. Most people work for mid- to large-size companies with established policies and procedures in place. In such environments, being innovative is much more challenging. Almost everyone who has tried a new idea in a traditional corporate setting and had the idea fail can tell sad stories of how the experience negatively affected their career.

Most companies are like legacy systems: they are resistant to change. Being creative here requires discipline. This may sound like an oxymoron, but innovation must be more incremental for the company to accept it. This is not necessarily the best way of doing things, but it is the way things are done.

Innovation in most companies is like security: it's only as strong as the weakest link in the chain. In my experience, lower-level managers are often more focused on keeping their departments running on a day-to-day basis than supporting the strategic direction of the organization. The results of innovation are often hard to measure, but the results of failure on mundane tasks can quickly become painfully obvious.

# # #

UBER

Technology disrupts the taxicab business

July 22, 2016

I finally got to try Uber last weekend. Uber is the poster child for disruptive technology in the taxicab space. Instead of hailing a cab and hoping for the best, you order an Uber car via their impressive smartphone app. It worked like a charm.

I was in New York City, wondering how I was going to get from LaGuardia airport in Queens at midnight to the hotel in midtown Manhattan. Just 10 miles separates these two locations, but in many ways they are worlds apart. While I was waiting for my luggage to arrive, it occurred to me that trying Uber might be preferable to taking a chance on the famous NYC cabs.

The Uber app shows you exactly where their cars are in real time, so you can see your options. You select a car and a time, and a confirmation appears. You know the type of car, the name of the driver, and even their phone number – which are really useful when it comes to finding your Uber at an airport.

Upon exiting the terminal, I was met with what I assume is a typical New York scene: barely organized chaos. There are several lanes, separately by concrete barriers, for cars, buses, and cabs. It was not clear which lane the Uber driver would use, so I called him. He answered right away, told me how many minutes before his arrival, and said that I should look for the silver 'U' light in his window, which indicates an active Uber car.

When he arrived, several other people moved towards the car. When I mentioned that this was the Uber I ordered, they said the same thing. Uber in NYC has a new default option of ride sharing, which the driver is obligated to follow. So my first Uber drive turned into an interesting social experience with a few marketing guys from Boca Raton.

Ride sharing does have one advantage: the cost is split among all passengers – and it's done automatically via the app. In fact, the entire payment is handled online through your Uber account. No cash, no credit cards. You don't even have to provide a tip right away: when you receive your email receipt, you have the opportunity to tip your driver. It takes a lot of pressure off of weary travelers.

You also are asked to provide a review of your Uber ride. Drivers rely on positive reviews for business. I found the Uber cars to be very clean and the drivers extremely courteous. Certainly it was a marked improvement from a traditional yellow cab in the Big Apple.

# # #

PHOENIX

Technology to beat the heat

July 29, 2016

PHOENIX, Ariz. – Living in Florida you grow accustomed to the heat. The abundant sunshine and generally pleasant weather is part of what makes Florida so attractive. But when July arrives, and the temperature and humidity both start to climb, you begin to wonder if it might be more comfortable to spend the summer elsewhere.

Many snowbirds head north, to cool mountains and refreshing lakes. I take the opposite approach: I head to Phoenix. Out here in the desert, where it was 114°F for several days in a row this week, you learn what true heat is like. When I return to Florida, the sticky weather doesn't seem so bad.

The greater Phoenix area, appropriately referred to as the Valley of the Sun, is home to over 4.5 million people. How do Phoenix residents cope with the extreme heat? How does the city manage to thrive in such harsh conditions? If average temperatures are indeed rising, what lessons does Phoenix hold for other urban areas?

The short answer to all three questions is the judicious use of technology. For example, I would not want to live here without air conditioning. The oppressive heat would make each summer day a misery. Cooling is actually more energy efficient than heating, so overall the utility bills for many people are lower here than they are for northerners. We can thank Willis Carrier for his invention over 100 years ago.

Water is an increasing concern in the southwest. There are no natural sources of water in the Phoenix area. In fact, according to the government, the Colorado River accounts for about half of the city's water supply, so a lot of infrastructure work has gone into making the city more desert-friendly. For example, older homes in the Scottsdale area have green lawns, which require a lot of watering. Back then, people wanted their home to resemble the homes they left back east. Now, almost all newer homes are xeriscaped: yards are creative layouts of rock and natural plants that don't require much watering. Most yards are crisscrossed with tiny black tubing to deliver water via computer control directly to where it's needed; you rarely see sprinklers sending water into the air indiscriminately.

Phoenix receives its power from a mix of coal, natural gas, nuclear, and hydroelectric sources. Since the city is the sunniest large metropolitan area in the country, there is a huge emphasis on solar energy. You can see solar panels almost everywhere, which aren't always the most visually appealing. New developments in smart materials may make buildings and roads active solar panels themselves, which would be truly innovative.

# # #

STAR TREK AT 50

What are your favorite Star Trek memories?

August 5, 2016

This year marks the 50th anniversary of the original "Star Trek" television series. When it debuted in 1966, many of the shows exciting new concepts were purely fictional: alien life forms, anti-matter power sources, energy weapons, faster-than-light travel, intelligent robots, portable communicators, and transporters all sprang from the fertile imaginations of the show's writers. But as with all good science fiction, the stories were grounded in science fact. Now, a half-century later, technology has made many of the show's fictional concepts a reality.

My first memories of "Star Trek" are of watching Captain Kirk battle a reptilian being from the Gorn species on a barren planet. The episode was called "Arena." I was too young to watch it when it was originally broadcast on January 19, 1967, but like millions of other budding young science aficionados, I watched it a few years later when the series went into reruns. Even on a tiny 13" black and white TV set, with a snowy signal and poor vertical hold, I looked forward to "Star Trek" each week.

My strongest impression of "Arena" was not the fancy gadgets that Kirk used or the technology found on the starship Enterprise. Instead, I was most fascinated with how Kirk was able to construct a crude weapon using a bamboo shaft, black powder, and diamonds he found lying around. I was in Boy Scouts at the time, so outdoor survival skills were much in my mind. Kirk overwhelmed the Gorn with his improvised rifle, but then refused to kill his opponent just to satisfy the demands of the powerful alien race called the Metrons who arranged the forced combat. In doing so, Kirk demonstrated the advanced concept of mercy and thereby saved himself, his crew, and his ship. This moral dimension of "Star Trek" was something that was an important part of the show's message of hope for the future.

Watching the television show fostered a lifelong interest in all things "Star Trek." I read all the pocket books from the original series. I even kept a handwritten summary of each story, much of which was written while at the beach in Maine over several summers.

On Friday, August 26, at 8:00pm in FIT's Olin Engineering Auditorium (Room 118) I will present a free lecture called "Star Trek at 50." This talk is part of the College of Science's Public Lecture Series. I encourage you to come hear more about the science, technology, and engineering of "Star Trek." Find out what is fact and what is fiction. Meanwhile, please feel free to drop me a line with your favorite "Star Trek" memories.

# # #

Star Trek at 50: Communication

Beam me up, Scotty!

August 12, 2016

"Beam me up, Scotty!"

This phrase, perhaps more than any other, epitomizes the original "Star Trek" television show of 50 years ago. It was never actually said, but Captain Kirk said something very similar to Chief Engineer Montgomery Scott (Scotty) on numerous occasions.

There were no cell phones in 1966, but the Communicator Kirk used was deeply influential on the design of the phones to follow 30 years later. The Motorola StarTAC, released in 1996, clearly resembled the Communicator from the 23rd century. Both devices fit in your pocket. To make a call, both devices were "flipped" open; snapping it shut ended calls. But there the similarities ended.

Kirk's Communicator didn't have a keypad; he never had to dial a number. Instead, he just spoke commands, such as "Kirk to Enterprise," and someone on the ship answered right away. In that sense, the Communicator was more like a walkie-talkie than a cellphone. Today we have similar voice activation technologies, such as Apple's Siri.

When Kirk used the Communicator, he was usually down on the planet's surface and Scotty was in orbit aboard the Enterprise. The International Space Station is in orbit nearly 250 miles above the Earth; that's a long way for a cell phone signal to travel. There were times when the starship was even farther away, but it was very rare for Kirk to be out of range; apparently there are few "Can you hear me now?" moments in deep space. Today's cell phones operate on regular radio bandwidths. The fictitious Communicator relied on subspace to overcome poor signal strength and vast distances.

Think of subspace as a version of Einstein's curved space-time. Communications used subspace to avoid traveling at the speed of light, which would take too long to support interactive communication. For example, it takes over 48 minutes for a signal from the Juno probe around Jupiter to reach Earth – hardly fast enough to support a Skype call. Subspace is a MacGuffin the writers used to get around this physics problem.

The Communicator didn't have video. It also didn't have any apps, in stark contrast to today's smartphones. I guess Kirk didn't need to update his Facebook status for each mission. But the Communicator did have something like a GPS transponder in it, which is how Scotty could lock onto the away team to beam them back to the ship.

The original handheld Communicators were eventually replaced with lapel badges in "Star Trek: The Next Generation." It's interesting that the real world has not followed suit. Our cell phones are still mostly handheld devices, with the exception of the Apple Watch.

# # #

STAR TREK AT 50: TRANSPORTATION

I cannae change the laws of physics!

August 19, 2016

"Ahead warp factor one, Mister Sulu."

This order was often given by Captain Kirk at the end of an episode in the original "Star Trek" television show. Captain Picard would later use a variation, "Engage!" in the follow-on series "Star Trek: The Next Generation." In both cases, the Enterprise captains were instructing the bridge crew to pilot the ship into the fictional world of faster-than-light travel.

The actual definition of warp speed in "Star Trek" changed over time, but warp one was usually defined as the speed of light, which is 670,616,629 mph. If the Enterprise traveled at the relatively pedestrian speed of warp one, it would take 4.37 years to reach our nearest start (other than the Sun), Alpha Centauri. Clearly that would severely limit the ship's mission to "boldly go where no one has gone before" to our immediate galactic neighborhood.

The fastest spacecraft NASA has built is the Juno probe to Jupiter. It was speeding along at 165,000 mph. When you consider that the average bullet travels at 1,700 mph, Juno was really moving quite fast: 97 times faster than the bullet. But that's still just 0.00025 times the speed of light – a veritable snail's pace in space.

The Enterprise can travel faster than warp one, but later editions of the canon limited the ship's top speed to warp ten. Other alien races could conceivably go faster using trans-warp technology. But back here on Earth, we're still searching for warp speed to take us to the stars.

One limitation is the engines used in today's rockets. They just don't generate enough energy to push a spaceship close to the speed of light. Chemical reactions are limited in the amount of energy they can produce, and atomic power has its own problems. We need something with more kick.

The Enterprise uses matter/anti-matter reactions to create a bubble in space-time to travel in subspace. This is fictional of course, but it's been a dream of scientists for a long time. Stephen Hawking has written extensively about black holes and how they may permit faster than light travel (and even time travel). But turning these theoretical ideas in practical engineering is something else entirely.

The movie "Star Trek: First Contact" focuses on human's first meeting with the Vulcan race. Contact was made possible by the first warp drive, developed by Zefram Cochrane in 2063. That means we still have nearly 50 years to wait before we can travel to deep space. Unless of course we invent something not dreamed of in science fiction – and sometimes, the real world is even more surprising than "Star Trek" envisioned.

# # #

STAR TREK AT 50: CIVILIZATION

Dammit Jim, I'm a doctor, not a bricklayer

August 26, 2016

"It's life, Jim, but not as we know it."

This line, commonly attributed to ship surgeon Dr. Leonard ("Bones") McCoy while speaking to Captain Kirk in the original "Star Trek" television show, was never actually said. A few variations were used, but it was science officer Spock who used them, not McCoy. Nevertheless, the meaning of this trope has come to symbolize the alien nature of the various life forms the Enterprise encountered during its five-year mission.

I am still surprised each time I see the Season Two episode called "Catspaw," when the aliens Sylvia and Korob, who have appeared human the whole time, are finally revealed to be Ornithoids – tiny squid-like creatures less than a foot tall. The effects were crude, but the impact was strong. It made me think about the nature of intelligent life, and I realized that a primary reason older science fiction shows represented aliens as minor variations on the human form had to do with actor makeup and production budgets. Plus, the audience can relate more to characters that look vaguely like them and are not *too* alien.

In the Season One episode called "The Devil in the Dark," the writers focused on what it means to be alive. The crew encounters what appears to be living rock, based on silicon instead of carbon, which has been attacking miners who are unknowingly destroying the alien's eggs. When the landing party realizes that the injured rock is alive, Kirk tells McCoy to treat it, to which McCoy replies, "I'm a doctor, not a bricklayer." It seems even the civilization of the future grapples with the essential meaning of "being human."

In addition to organic forms of life, robots also appear in several "Star Trek" episodes, but they are aliens. Some of the robots are crude, while a few appear to be sentient. It's not until much later that the android Data appears as a full-fledged member of the crew in "Star Trek: The Next Generation."

In the "Star Trek" world of the future, the galaxy turns out to be full of beings, some friendly and some hostile. If I were a betting man, I'd say our best chance of detecting life (however we define it) in the real world would be in the subsurface ocean on Europa, a moon of Jupiter. But maybe NASA probes will uncover signs of life on Mars. Maybe the tin-foil hat brigade is right and the aliens are already here on Earth, living among us. No matter where we find it, the proof that we are not alone in the universe would utterly change our civilization forever.

# # #

IMAGINE THERE WAS NO LABOR

Life in a post-jobs society

September 2, 2016

Most of us dream of the day when we don't have to work. You spouse may have snide remarks that suggest the day has already arrived. But the truth is, we are working longer hours than ever before – and you can blame technology in large part for your predicament.

Our always on, always connected modern world lets us work from almost anywhere. But it also blurs the line between work and play. For many professionals, 9-5 days are a thing of the past. Technology has given us a lot of freedom – including the freedom to work 24/7/365. And even if you think going off the grid for occasional downtime is a good idea, your boss may think differently.

This digital drudgery may soon change, however, by further advances in technology. Improvements in artificial intelligence software and robotic hardware will enable machines to perform the duties of many white-collar workers. Pretty soon you may be able to imagine there was no labor – if you dare.

In the television series "Star Trek: The Next Generation," the 24th century is portrayed as a worker's paradise. The economy of the future is vastly different. For example, they have done away with currency. People aren't paid for their work; instead, they work solely for the satisfaction of bettering themselves. Most people in Captain Picard's world focus on creative endeavors; every new world they visit seems to be some sort of artist colony.

In this utopian society, people have been freed from most forms of manual labor through the widespread use of autonomous devices. Those who still work with their hands tend to be craftsman, not ditch diggers. But would this really be such a good thing?

I'm not sure we're ready for a post-job society. Consider the problems that massive unemployment causes in the world today. There are too many young people with too much time on their hands, unprepared to live in such an unstructured manner. And there are too many older people who have lost the traditional reason for getting up in the morning. There are only so many video games you can play, or daytime TV shows you can watch, before you become bored and restless for something meaningful to do. But our education system has not prepared most people to think creatively and to act independently in a way that is beneficial for themselves and for the greater good.

For now, we're more Ferengi than human when it comes to the accumulation of personal wealth. We follow our own rules of acquisition. Technology may advance by leaps and bounds, but I don't foresee Gordon Gecko retiring anytime soon.

# # #

RIP HEADPHONE JACK

Oops, they've done it again

September 9, 2016

They've done it again. Apple is well known for killing off technology that they see as part of the past. They famously stopped shipping computers with floppy discs many years ago. More recently, they stopped shipping notebook computers with built-in CD/DVD drives. Sometimes they even abandon technology that they helped create and promote, such as FireWire.

Apple is also famous for introducing new connectors on their products. For example, the iPod has used several pin configurations since it was originally released. Each change in connectors forces users to either upgrade to a new device, or use a dongle adapter to connect the old to the new – never an elegant solution.

This week Apple announced the iPhone 7. Their new smartphone has a number of improvements, but arguably the most controversial change with the new model was the elimination of the headphone jack. Instead of the tried-and-true 3.5mm port, Apple is forcing its customers to use new headphones that use a Lightning connector (the same connector used for power), or to use wireless headphones that don't need any physical connection. Apple is also providing a Lightning to 3.5mm adapter for those customers who don't want to ditch their current headphone investments.

The 3.5mm connector is ubiquitous. It's been around for over 100 years, and about 50 years in its current form. There are literally billions of devices that rely on it. Nevertheless, Apple views it as yesterday's technology, and has made the courageous decision to eliminate it from the iPhone.

Apple has given several reasons for their decision. One reason is size: the new iPhones are thinner than ever, and the physical diameter of the 3.5mm plug was a limiting factor in how thin they could make the phone's enclosure. Another reason is dirt and water: by removing the 3.5mm port, they close a gaping hole in the phone that water easily enters and ruins the inner electronics. A third reason is audio fidelity: the old headphone jacks were analog, which cause loss of signal purity when the original sources (music, voices) are converted from their native digital format.

Apple is selling a new pair of wireless headphones, called AirPods, which look very slick. One of the problems I've always had with Bluetooth devices, including headphones, is the difficulty of pairing them with the phone. Apple says they have addressed this with a seamless pairing process that requires the user to just press "Connect" on the phone once. The new version of Bluetooth that will be released soon should make things ever better.

But you'll have to shell out an additional $159 to enjoy your new AirPods.

# # #

Johnny Cab Arrives in Steel City

The future of transportation takes on Pittsburgh's roadways

September 16, 2016

I'm one of those people who almost never get lost. While driving in a city like Los Angeles, I'm amazed that I can still find my way around without a map or a GPS system – even though I've not lived there for nearly 15 years. The same thing happens when I visit a new city for the first time: a quick glance at the map to get my bearings, and I'm generally good to go. That even includes places like Beijing and Munich, where the street signs are not even in English.

But there's one city that has always baffled me: my old hometown of Pittsburgh.

I know that other cities have their driving challenges. The myriad one-way streets of Boston. The crazy congestion of New York City. Driving on the other side of the road in London and Sydney. But to me, nothing compares to the challenge of trying to drive around the city that Mellon built.

I think the directional problems started the first day I was in Pittsburgh. I was working at Carnegie Mellon University (CMU), and as part of my relocation package they provided me with map of the campus and of surrounding areas. The diagram itself was clear enough – except that north was marked with an 'N' pointing down at the bottom of the map.

Who puts north pointing down?

My internal compass never recovered.

75

Driving into town from the northern suburbs was an exercise in crazy loops; there never seemed to be a straight line from A to B that you could follow. Driving out of town was a dangerous game of rapid lane changes to avoid getting locked into a path that would take you across one of Pittsburgh's many bridges and/or through one of its many tunnels. Since the signs were all but useless, I never quite knew where I was headed. And asking long-time residents for directions was also frustrating, since they always referred to old landmarks that had long-since disappeared.

So, imagine my surprise to learn that Uber has chosen Pittsburgh as the first city to deploy a fleet of driverless cars. Uber has partnered with CMU in this project, leveraging the university's impressive robotics expertise to help the cars navigate the Golden Triangle. In my opinion, they couldn't have chosen a more challenging location to begin their trials. Good luck with the North Hills and the Strip District!

It will be interesting to see how Pittsburghers react when they hail a cab to the next Steeler's game, and instead of a human cabbie they are met with Johnny Cab from "Total Recall."

# # #

STAYING FOCUSED

Take the slow boat to increased productivity

September 23, 2016

In today's media-rich world we are constantly bombarded with digital interruptions. Instant messages, tweets, Facebook updates, email notifications, and even the occasional phone call all vie for our attention. It's no wonder that many people find it difficult to stay focused.

The younger generation seems to thrive in a multi-tasking environment, but I question whether or not they are truly making efficient use of their time. What appears to be multi-tasking is often task switching, and the latter is notorious for reducing productivity. In the computer world, we say that an operating system that is switching tasks too frequently is thrashing, and that's what I see many folks doing every day.

Fortunately, for every modern problem, there's a proffered solution. In this case, there's literally an app for that. There are apps that block websites, so you won't be tempted to check the latest sports scores instead of finishing your report. There are apps that block all messages and notifications, so you're not seeing updates popup and clutter your screen all the time.

It says something that we need a software program to keep our attention focused on another software program by blocking a third set of software programs. It's an open question why you could not simply turn these programs off yourself, but hey, that's what the apps are for.

Writers are notorious for being easily distracted. I've tried a program called OmmWriter that offers a vastly different experience than Microsoft Word. OmmWriter takes over the whole screen. There are no menus, just a small textbox to enter your writing. The background is a rather bleak winters scene, all white with a few leafless trees in the distance. Tranquil music plays. When you type, the keys make reassuring squishy sounds, like water dripping. Some authors swear by it.

I may make fun of some of these apps, but I'm guilty of using other methods to stay focused. Years ago I would work wearing headphones with loud rock music blaring in my ears. Now I find the racket distracting. I admit that I'm not above putting on the occasional ambient music track in the background.

But to fully realize the lengths some people will go to stay focused, consider the "Slow TV" channel from Norway (now available on Netflix). It's a collection of long movies that capture mundane events in their entirety. For example, there's an eight-hour show that records a train journey to Oslo. The show is just the view from the front of the train, with the occasional announcement from the conductor. That's it. For eight hours.

It's quite good actually.

# # #

MUSK'S MISSION TO MARS

Motive. Means. Opportunity. Let's go!

September 30, 2016

Buzz Aldrin has a clear motive for getting to Mars in a hurry: July 20, 2019 is the 50th anniversary of his Apollo 11 mission to the moon. The year 2040 had previously been mentioned as a possible timeframe for a manned mission to the red planet, but politics and technology make such predictions very fluid. Whatever the date, there's an emerging consensus that Mars should be our next destination for human exploration.

This week, Elon Musk unveiled an audacious plan to have us colonizing Mars within a decade. His company, SpaceX, will provide the means for us to do so: the Interplanetary Transportation System. Musk is proposing a new series of large, modern, and reusable rockets that will ferry astronauts into Earth orbit, return to the launch site to pick up more fuel, and return to orbit to join the spaceship for a continuing journey to Mars. SpaceX has already demonstrated it can stick landings of their current rockets, so it's not too difficult to imagine them taking this approach to the next level.

The Interplanetary Transportation System still has a number of significant technical challenges. For example, the engines on the current Falcon 9 rocket are not powerful enough to get a ship to Mars. But SpaceX is already testing their next engines, called Raptor, which will be used on the Mars spacecraft – 42 of them at once, in fact. Ultimately, we'll need something even faster to get beyond Mars.

There are also issues related to radiation, food, and the effects of long-term isolation on humans living off world. We've been studying these problems for years, but we'll never really know how to solve them until we actually experience something like a trip to Mars (and back). The movie "The Martian" was full of amazing science, but a lot of it was still "CSI"-like fictional.

Musk's vision is to provide the opportunity for all of us to participate in space travel and ultimately experience life on another planet. NASA has led the way towards this goal for decades, and it continues to play an extremely important part in the overall mission to explore our solar neighborhood. But there's a clear increase in the role of private enterprise when it comes to space, and that's probably a good thing. We need thought leaders like Musk, Bezos, and Branson to provide inspiration to the next generation of scientists and engineers to get involved in this amazing race.

Motive. Means. Opportunity. It seems like we have everything we need to commit the perfect caper: be the first humans to colonize Mars. As Musk said, "Let's make life interplanetary!"

# # #

STEVE JOBS

He's been gone five years. Wither Apple?

October 7, 2016

Steve Jobs, the co-founder of Apple (and other companies), passed away on October 5, 2011. He succumbed to pancreatic cancer at the age of 56. I wrote about his death in this column exactly five years ago.

Since Jobs' untimely demise, Tim Cook has led Apple to continued success. He has done a very competent job of managing the company, primarily by releasing incremental improvements in Apple's impressive product line. The recent release of the iPhone 7 is just the latest example of this evolution.

But has Apple lost its edge?

Where are the truly new and innovative products? Where are the revolutionary designs that make people want to line up outside the Apple store all night to get their hands on the latest shiny gadget? Where is the visionary leadership?

In the book "On the Firing Line: My 500 Days at Apple," Gil Amelio wrote about his experience taking over as Apple CEO from Michael Spindler after Steve Jobs had been fired. Amelio's background included more staid and traditional companies such as National Semiconductor. When he began his tenure at Apple, Amelio was surprised to find a dysfunctional company that had drifted away from its initial mission and its customer-centric focus. He lasted less than two years at Apple before being replaced in a virtual boardroom coup by Steve Jobs.

Jobs' second period leading Apple was even more successful than the first. But the period between his forced departure and his triumphant return was marked by inner turmoil, lackluster products, and a sagging stock price. It seems there really was something to Jobs' vaunted "distortion field" that had a positive effect on the company as a whole, even if he was quite hard on his employees.

We are now in the second go-around of Apple without Jobs. There has not been as much market confusion as the first time (e.g., selling Mac clones), but there has not been very much new in the product line either. It rather feels like Cook is just coasting, relying on the prestige of the past. But there's a limit to how long this can last and still maintain a leadership position, and I think Apple is getting very close to it.

Apple is reported to be working on several new products. These include a virtual reality / augmented reality device, possibly a connected TV, and of course the rumored Apple car – electric, self-driving, and totally immersive. But they are chasing Tesla (among others) in this market. Time will tell if new leadership is needed to lift Apple to the next level. Is Elon Musk busy?

# # #

SAMSUNG GALAXY NOTE 7

Samsung has a meltdown. Literally.

October 14, 2016

On a recent flight to Raleigh I was surprised to hear the flight attendants add an additional warning to their usual spiel prior to liftoff. They admonished all passengers with a Samsung Galaxy Note 7 smartphone to avoid using (or charging) the device for the entire duration of the journey. That was the first time I can remember the FAA calling out a specific manufacturer for a market-bruising tongue-lashing.

The Galaxy Note 7 has the unfortunate characteristic of catching fire. The problem appears to be an issue with the batteries. I'm sure Samsung thought they had a hot product when it was first released just a few months ago, but I'm equally sure this was not what they had in mind. There's hot and then there's hot.

The negative publicity that has hammered Samsung has been brutal, and it couldn't have come at a worse time. They were getting closer to Apple in terms of innovative features, which led to increased public awareness of the brand as a viable alternative to the iPhone. Now Samsung has been set back worldwide.

The whole meltdown fiasco was poorly handled from the beginning. Samsung was late to acknowledge the problem. Then they were late to issue warnings and recall notices, leaving it to the major carriers to remove the phones from their shelves. This week Samsung announced the Galaxy Note 7 was being discontinued entirely. Just in time for the holidays too.

Apple's stock has risen considerably while Samsung has lost billions in market value. For Apple, the Galaxy Note 7 has been the gift that keeps on giving. For Samsung's customers, it's been a rude awakening. This is not the first time phones have been recalled due to battery issues, or even the largest recall in the smartphone industry, but it's certainly one of the most damaging.

Rubbing more salt in Samsung's wounds has been the media's reaction to Samsung's stated policy on the defective Galaxy Note 7s that customers are returning. Samsung has said they will not refurbish or repair any of the phones. Instead, they will dispose of them. Depending on how the disposal is carried out, the environmental impact of dumping several million smartphones full of rare earth elements back into the earth may have significant negative consequences. Even if Samsung does implement the disposal in a sustainable manner, their lack of clarity has created worse PR they didn't need.

Remember that the next time the flight attendants tell you that there's no smoking on the flight, don't roll your eyes and mutter than you already know that; they may be referring to your phone.

# # #

PROGRAMMING LANGUAGES

From simple to complex

October 21, 2016

On October 15, 1956, about 60 years ago, IBM introduced the programming language FORTRAN to the world. The word "FORTRAN" stands for "formula translation." FORTRAN was the first widely used high-level programming language. Prior to FORTRAN, programmers had to code in assembly language, a very low-level representation that closely mirrors the instruction set of the underlying central processing unit. Very few programmers use assembly language today.

The FORTRAN language was designed to "closely resemble the ordinary language of mathematics," which was appropriate considering its first use was for scientific computing. FORTRAN has been updated several times over the decades, but it remains true to its original purpose. Although FORTRAN is not taught in many computer science departments anymore, it's still the language of choice for work in areas such as experimental physics and high-performance computing.

The FORTRAN manual referred to the "automatic coding system" for the IBM 704 mainframe computer. The manual is an excellent example of brevity and clarity. In just 50 pages the entire FORTRAN language is clearly explained, along with coding examples and instructions on how to run the compiler that shipped the following year.

Skip forward nearly thirty years, and on October 14, 1985, Bjarne

Stroustrup of AT&T Bell Laboratories published the first C++ reference guide. C++ was a new programming language that extended the popular C programming language with object-oriented features. Like the first FORTRAN manual, the first C++ manual was about 50 pages long, although it relied heavily on programmers' knowledge of C to explain C++'s new features.

The first programming language I learned at school was Pascal. I then learned C a few years later. When I worked at IBM, I developed compilers for C and C++, which gave me deep insights into the languages and their design. I learned to appreciate the simple constructs of C when compared to older languages like PL/I. This is important, because too many of today's students are turned off by the seemingly incomprehensible syntax of some modern programming languages.

I recently examined the latest draft of the next C++ revision. It's huge. Books that explain how to program in C++ have become veritable tomes. It seems the days of a 50-page reference manual are over.

In my opinion, there have been so many dubious additions to the C++ language that it makes programs very hard to understand. Complexity is the bane of engineering, so it saddens me to see languages like C++ (and Java) evolve into unwieldy beasts than only experts can master. It's contrary to the goals of encouraging more young people to adopt a career in computing – and that's a pity.

# # #

TERRIFIED OF BAD RATINGS

What would you be willing to do for a five-star review?

October 28, 2016

Forget about ghosts and goblins. If you really want to be terrified this Halloween, imagine a future where everything is based on user ratings and customer feedback – including your standing in society. That's the premise of "Nosedive," the first episode of the third season of "Black Mirror," an excellent television show that focuses on the role of technology in the near future.

We already rely on user ratings when it comes to picking a hotel or choosing which movie to watch. Feedback from fellow customers has trumped the value of advertising in many cases: we trust what other people say about their experience more than we trust the company selling the goods or services. "Nosedive" just takes this reliance to the next level, where everyone reviews everyone else all the time. Each person walks around with a rating floating in front of his or her face, like a book review on Amazon.com. Depending on the rating, which is changing all the time, citizens' interactions are limited.

This is not so farfetched. We already are moving towards a review-based economy in many areas. Consider Uber drivers. Their likelihood of being selected by a customer is partly dependent on the driver's online reviews, which incentivizes the drivers to make the customer's experience as positive as possible.

It's easy to see how such a system can be abused. Customers who post bad reviews online often do it for bizarre reasons, but the review's negativity still impacts the service provider. I often buy things online and sort the products based on five-star and four-star reviews; I ignore anything less. But I don't take the time to see why some of the reviews may have been three stars; the damage has already been done, and in general it's irreparable.

Now consider how such a rating system would affect your life. Imagine everyone you interact with is rating you all the time. They are posting their comments about you to social media sites in real time. People read these comments and form their own opinions about you without having ever met you. Would you be more likely to start treating everyone, including your family and friends, more like customers that must be made happy all the time? What would you do to rectify bad reviews? And what if you chose to go off the grid and avoid the review system completely – would this exclude you from social activities? Jobs? Benefits?

This dystopian view of the future is not too far off. Just watch how many people walk around like zombies now, staring at their smartphones, posting constantly. Be afraid, very afraid.

# # #

ELECTION 2016

It's Florida: Vote early and often!

November 4, 2016

Baseball aficionados are inundated with seemingly trivial statistics about every aspect of player performance, the dynamics between team members and the opposition, and even the likelihood of success for specific scenarios (e.g., getting a hit with two runners on base in the fourth inning against a left-handed pitcher). Slow airtime between innings is filled with sports analysts droning on about this sort of minutia. The technology underneath the analysis is impressive: big data, predictive modeling, and machine learning all play a role in turning gigabytes of raw numbers into information viewers may be interested in hearing and seeing.

And yet, with all this computing power, almost no one predicted that the Chicago Cubs would win the World Series this year. And why would they? History suggested the Cubs would lose again, since they've lost every other year since 1908. Sometimes past performance is an accurate predictor of future events, but not always. In the business intelligence world, it's those events that don't fit the patterns that are the most interesting.

Which leads us to the 2016 presidential elections. Politics may have surpassed professional sports in its reliance on data models to predict outcomes. Polling has been around for a long time, so one might think that would make the accuracy of their predictions better and better over time. There certainly is more data to use as input to sophisticated models than ever before. We're constantly assured that numerous polls all point to a consistent outcome.

So why did Brexit happen?

When the United Kingdom voted in June to leave the European Union, the result was not at all what the polls had suggested. The majority of news services predicted the Remain side would win comfortably. Instead, the Leave side won, leaving pollsters scrambling to explain how they could have been so wrong. Again.

The pollsters may have had big data on their side. They may have had multiple models all showing the same voting patterns. In other words, they had technology on their side.

What they didn't have on their side was the human element.

Given the toxic atmosphere that surrounded the Brexit vote, many people lied when asked what they would vote for. I think something similar is happening on our side on the pond with the presidential election. Which means I will not at all be surprised if the final results are not in line with what all the polls are predicting.

Yogi Berra once said, "It's tough to make predictions, especially about the future." I trust this maxim more than I trust any pollster today. And that applies to baseball as well as politics.

# # #

WATCHING ELECTION 2016

I didn't watch the results – I surfed them

November 11, 2016

Like millions of Americans this week, I was mesmerized by the presidential election on Tuesday night. But I didn't watch the election coverage on television; I watched it online. It was the first time that I was informed of results by tweets, posts, and talking heads that I didn't even know existed, rather than watching a traditional news broadcast. It was a fascinating experience.

Since I lack cable television at home, most of the major networks are not an option for me. They do have apps you can use to watch their content, but most of the apps require you to also have a cable subscription, which I think rather defeats the purpose. I could have watched some coverage by over-the-air antenna, but I wanted to see what was available from the new media.

The answer was: a lot. Not all of it good, but certainly different. Buzz Feed anyone?

For example, I watched the last two presidential debates live. But Twitter of all things streamed the feed. I thought Twitter was all about short messages full of hash tags and other cryptic syntax. I didn't know they also do video streaming now too. I don't know how that supports their business model, but it did help me watch the debates in HD for free.

Many news websites, including FLORIDA TODAY, unlocked their content for the election. This meant I could watch CNN or Fox

News live, but only on my computer. With a bit of work you can cast the video from the computer to your TV, as long as you have the right hardware/software setup. I do this sometimes, but for the election coverage I mostly kept within the browser.

In fact, my browser was a veritable forest of tabs, each open to a different news source. I found myself skipping across different online channels, live streams, and social media commentary all night. It felt more participatory than just sitting and passively watching the analysts fill in the dead air while they waited to call another state.

My phone was also constantly beeping. Tweets from news sources sent me "Breaking News" alerts every few minutes. Friends shared their comments on developments as the night progressed. Even the occasional phone call added to the mix of severe data overload.

It was only four years ago, during the 2012 election cycle, that I was pretty much like everyone else when it came to watching the results: glued to the TV. Now, I rely on new technologies and new media to supply the data I need. It's messier, but it's also more immediate. And who has time to wait?

# # #

AI TREATIES

Is a "Strategic AI Limitation Treaty" needed?

November 18, 2016

At this week's Big Data Florida meeting there was an earnest discussion about the potential dangers of artificial intelligence (AI) run amok. Many of the world's leading thinkers have expressed concern about the extremely rapid advancements that are being made in AI research. Stephen Hawking has been quoted as saying that true AI could be humankind's most impressive invention – and our last.

It seems inevitable that AI's march towards increased capabilities will continue. Therefore, it is prudent to plan for a time (perhaps in the not-too-distant future) when strong AI becomes a reality. As with all technologies, AI could be used for good or for evil. Or it could take matters into its own hands (metaphorically speaking) and make decisions outside the normal ethical bounds in which society operates.

A suitable analogy to the possible worldwide disruption caused by AI is the proliferation of nuclear weapons after World War II. History bears witness to the incredible destructive power of thermonuclear devices. They've been used. We know the damage they can cause. For these reasons, the world's nuclear powers have striven to limit the proliferation of atomic bombs. Various treaties, such as SALT I and II (Strategic Arms Limitation Talks) codified bilateral agreements related to nuclear weapons between the Unites States and the Soviet Union.

We've seen how these treaties, along with a policy of "mutually assured destruction," have helped keep the nuclear peace for decades. There are still ongoing issues related to rogue nations and illicit trades, but for the most part these mechanisms have worked.

There are no such treaties in place for AI. And we have no way of limiting access to the software that powers AI. Their emergence would represent a singular event.

With AI, there are no exotic materials needed: no plutonium or enriched uranium. No special centrifuges. No ballistic missile delivery vehicles. All that's needed is a copy of the source code and a computer network. After that, AI could spiral out of control exponentially faster than the spread of nuclear weapons. At the moment we're ill-prepared for the consequences of such a disruptive moment.

Perhaps the world needs a new SALT: "Strategic AI Limitation Treaty." This treaty would involve more than just the cold war superpowers; it should involve at a minimum the G20 nations to begin with. Controlling the spread of AI will be considerably more difficult than the spread of nuclear weapons, but it's something we need to start thinking about.

Nukes don't take over their masters, but AI might, which makes AI even more worrisome. If we don't act soon, the decision might be taken away from us.

# # #

FRIENDS

Be thankful for the friends in your life – digital or otherwise

November 25, 2016

Thanksgiving is many things to many people. For some, it's all about shopping and Black Friday deals. For sports fans, it's a day of binge watching football. With nearly 50 million turkeys eaten over the next few days, food is clearly a big part of the celebration for a great swathe of the population.

Personally, I think the focus of Thanksgiving should be on the meaning of the holiday's title: giving thanks. And we all have a lot to be thankful for this year.

Technology's inexorable drive towards increased hardware and software capabilities has created some truly incredible new devices. Last year I wrote about Alexa, the ghost-like disembodied voice that lives inside Amazon.com's Echo personal assistant. Since that time, Alexa has gained some impressive new skills. The recent introduction of a diminutive version of Alexa in the Echo dot has just increased her usefulness. I am thankful that Alexa is the first product that many technology-phobic seniors can use without any special training: they just talk to her. Indeed, for many people living alone, Alexa has become an always-on companion to chat with whenever needed.

This year I welcomed Watson into my life. He's a Golden Retriever puppy that just turned eight months old this week. In the five months since I've had him, my life has changed significantly. Not only is my daily routine now very different than it used to be (e.g., walks in the morning and potty breaks for him in the middle of the night), it's much richer as well. I'm forced to step away from the computer at regular intervals because he wants to play toss or have his belly rubbed. At first I found his interruptions annoying; now I look forward to them. His wet nose and slobbery kisses are a healthy reminder that there's a lot more to life than just work. He's a refreshing technology-free zone.

This Thanksgiving, remember what an incomplete angel named Clarence once told a despondent banker named George, "No man is a failure who has friends." I am president of the Space Coast Writers' Guild and editor of our new anthology, *Friends*, which is a collection of stories about friendship in all its forms. Whether your friends are people (real or online), diamonds, or dogs, friendship is the basis for a happy and successful life.

Throughout our lives and our careers, friends come and go, but I'm fortunate to still be in contact with some of them from my childhood. That's me in the photo, taken in 1969, with my best friend Richard. I'm the one on the right with the awesome haircut.

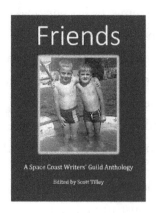

# # #

THE ULTIMATE COMPUTER

Will we all become dunsels?

December 2, 2016

Season 2, episode 24 of the original "Star Trek" television show was called "The Ultimate Computer." It was broadcast on March 8, 1968 – nearly a half-century ago. It was incredibly prescient in forecasting the potential dangers that intelligent machines represent.

In the show, a new computer called the M-5 is given a trial run on board the Enterprise. The M-5 is so powerful that it can operate the ship all by itself, effectively making the crew redundant. Captain Kirk is not happy with the situation, but Science Officer Spock is fascinated with the computer's capabilities. Star Fleet agrees to the experiment because Richard Daystrom, who had invented the advanced computer systems already used on starships, also designed the M-5.

To imbue the M-5 with artificial intelligence, Daystrom programmed the computer with his own memories and abilities. In effect, the M-5's software was based upon Daystrom's brain. The unexpected result was that the M-5 quickly developed a sense of identity and self-preservation. It stopped listening to commands given by humans and took actions that resulted in widespread death and destruction. Kirk eventually gets the M-5 to shut itself down by arguing that the computer is working against its purpose to "save men from the dangerous activities of space exploration."

As with many things from "Star Trek," science fiction is now becoming science fact. We already have drones that can operate independently. Aircraft have cruise control and automated takeoff and landing capabilities. Soon our roads may be crowded with self-driving cars. All of these machines rely on programming and robotics to operate autonomously. For now, we provide the programming and we develop the hardware, but it's not farfetched to imagine a time when the machines improve upon their designs and create new versions of themselves – essentially evolving outside of our control. The irony is that we're working feverishly to make this happen.

The M-5 may have been flawed, but it did demonstrate that a machine could control something as complicated as a space ship. We do this today with far-flung probes and the rovers on Mars. This means a machine can perform the essential role of the entire crew, something that should be rather unsettling to us all. In fact, Kirk is mockingly called "Captain Dunsail" at the start of the M-5 trial. A "dunsel" is a part that serves no useful purpose. Ouch.

On Saturday, December 3, at 2:00pm in the Henegar Center (www.henegar.org) I will be talking about "Star Trek at 50." Please come to hear more about how "Star Trek" has profoundly influenced our communication, transportation, and civilization – now and into the future.

#

BIG DATA IN 2016: A YEAR IN REVIEW

Sophisticated machine learning goes mainstream

December 9, 2016

When I logged into Facebook recently I was presented with a video titled "Your Year in Review." The message accompanying the video read, "Scott, we made a video for you to look back at some moments from 2016. From all of us at Facebook, we hope you enjoy it!" Naturally, curiosity got the better of me, so I clicked on the video and watched it.

It was actually pretty good. The video is a collage of photos you've posted, pages you've liked, and posts you've made in the past year. You're even able to edit the video to replace the default selections made by Facebook's algorithms. Interestingly, the video was not curated by Facebook's minions carefully trolling through your social media activity and manually picking out the cutest cat pictures: it was created automatically by Facebook's algorithms. With over 1.5 billion users, how could it be otherwise?

The Facebook video is a great example of where big data has gone in 2016. This time last year, big data was just a twinkle in the eyes of savvy quants who believed there was value hidden in the petabytes of information many enterprises now produce, just waiting be unlocked. Twelve months later, companies like Facebook have capitalized on data science by making the complex seem simple. It's a sign of maturity when computing technology fades into the background, letting businesses focus on the user experience instead.

Underlying the advances made in big data during 2016 are rapid changes in the software programs that perform data mining and pattern recognition at scale. Very sophisticated techniques for sifting through massive amounts of data in near real-time have been codified in packages that programmers can use without worrying about implementation details. Google's TensorFlow is a good example: it is an open source library for machine learning that makes big data techniques accessible to non-experts.

There have also been notable advancements made in the hardware infrastructure needed to perform massively parallel computations. It was not too long ago that using graphical processing units (GPU) to perform numerical analysis was only possible for special-purpose applications. Now the power of GPUs can be harnessed for even the most mundane scientific calculations – and again without the programmer needing to know the intricate details of the supporting infrastructure.

Storage capabilities also continue to increase. Consumers can now purchase a 4TB drive for just over a hundred dollars. That's 4,000 billion bytes of data. However, according to Gwava, over 400 hours of video is uploaded to YouTube every minute of every day. So I guess you still can't download all the cat videos from Facebook just yet.

# # #

A POST-ALPHABET WORLD

We need a modern Rosetta Stone

December 16, 2016

This week in New York City, Donald Trump met with leaders of the top technology companies, including Amazon.com, Apple, Facebook, Google, IBM, Microsoft, and Tesla. Notably absent was Twitter. The President-elect's team said Twitter was not invited because it was too small compared to the other tech giants. Some pundits posited that it was because Twitter refused to create a "Crooked Hilary" emoji during the presidential campaign.

An emoji is an icon, like a smiley face, that is often used in short messages and tweets. The latest update to iOS contains a veritable library of emoji, which many people use instead of actual text to convey meaning. Emoji are particularly popular with young people – so much so that there's been some concern that we're entering a post-alphabet world.

It seems almost quaint that there's been a discussion about learning cursive writing in school. Personally, I'm fully in favor of it, but some educators (and administrators) feel that learning cursive is akin to learning how to shoe horses: an interesting activity, but an archaic skill not needed in today's world. The argument against cursive writing is that students will spend most of their time typing, not writing by hand with a pencil or pen, so why bother?

The unexpected popularity of emoji-based messages has made the argument moot. Forget about typing and text; just learn which icons to use. Even Internet jargon, such as the acronym "LOL," has become passé.

The use of emojis have become so commonplace that some professors become alarmed when their students send them text messages that are not replete with cartoon cats and thumbs-up signs. One linguistic anthropologist at the University of Toronto says that the use of emojis may replace text-only communication for the next generation of students – and for the better. We'll see.

In some ways it feels like we're moving backwards, not forwards, in terms of communication. The ancient Egyptians used hieroglyphs to write down their history. We appear to be dumping hundreds of years of modern language for pictures and symbols.

You can buy books (presumably written with actual words) on how to "speak" emoji. There are beta versions of search engines that let you look for content online using only emojis. There's even an entire version of the novel "Moby Dick" that has been translated into emojis. It's called, appropriately enough, "Emoji Dick." Really.

The automated translation systems don't yet handle emojis, but I'd welcome a simple emoji to English translator for my phone. I often feel the need for an updated version of the Rosetta Stone to understand what people are saying to me. ☺

# # #

ALEXA

Amazon.com's Echo Dot is the best product of 2016

December 23, 2016

I first wrote about Alexa a year ago and I've been relying on her ever since. Alexa is the artificial intelligence inside Amazon.com's Echo and Echo dot. The Echo is the original model; it looks like a black Pringle's can. The dot is the newer model; it looks like a hockey puck. The dot costs just $50 and is being marketed by Amazon.com as something you should buy in six packs – one for every room in the house.

The surest sign of Alexa's usefulness is that it passed the "Joyce test." This is like the Turing test, except that Joyce is my mother and she has a unique talent for immediately uncovering design flaws and operational bugs in new products.

For example, last year I tried to get her to switch to an iPhone and it was a disaster. Many of the things I'd tell her to do, such as "swipe left," were foreign actions. She doesn't have to "swipe" anything else she uses, so why should she have to swipe the phone just to make a call? Apple should hire her as a user experience consultant.

I've had similar situations with her and other electronic devices in the past, including computers, television remote controls, and even complicated radios. She always seems to do the "wrong" thing, but it's only "wrong" in the sense that the product's designer never thought anyone would do such a thing. However, that's what testers are supposed to catch.

With Alexa, my mom has had no such problems. She's been using an Echo dot for over a month now, and she thinks it's great. I firmly believe the voice interface is what has made all the difference. No black buttons on black backgrounds to push. No screens to swipe. Just talk and it responds.

My mom has particularly enjoyed listening to holiday classics. I setup her Alexa to have access to the Amazon.com Prime Music library, so she can stream literally millions of songs. She no longer has to fiddle with an old CD machine; she just asks Alexa to play one of her old favorites, like Perry Como, and voila!, the old crooner magically appears.

Alexa also let's my mom listen to radio stations that have weak broadcast signals. She can hear the news from home without worrying about where in the house she has to place the radio and which direction to point the antenna.

The Echo dot is my choice for best product of 2016. It's affordable enough to make a great gift for those people in your life who just want gadgets to work. And it does.

# # #

LOOKING BACK AT 2016

Artificial intelligence, big data, and cybersecurity

December 30, 2016

Merriam-Webster's word of the year for 2016 was "surreal." According to the online dictionary, surreal is "… used to express a reaction to something shocking or surprising." Surreal is also a good word to describe some of the key events in the technology world that took place this past year.

Artificial Intelligence (AI): The extremely rapid advancements that have been made in AI during 2016 were stunning. Research into AI is 50 years old, but only recently have new algorithms and new computing power come together to enable truly innovative applications. From a technical point of view, AI implementations have mostly switched from a structured and rule-oriented world to one that is data-driven and relies on machine learning. Whatever the internal mechanism, the result for the user has been remarkable.

Leading the way is Alexa, the AI embedded in Amazon.com's Echo products. Alexa's voice recognition is not perfect, but I find it far better than anything else available. More importantly, Alexa's capabilities grow almost daily, and in a way you hardly notice. To me, that's the sign of good design and sophistication: the computing fades into the background and you're left with just the device that responds almost like a real person. It's surreal.

Big Data: The interest in big data continued to grow in 2016. Marketing companies love the insight it provides them into consumer demand. Consumers remain blissfully unaware of how much personal information they are providing while online to anyone willing to vacuum it up – which gets easier to do every year.

The surreal aspect of big data this year was the spectacular failure of pollsters to get elections right. Brexit in the UK and the presidential election here at home both turned out quite different from what we were told they would. But the advanced models used in predictive analytics didn't fail. Rather, they produced the expected results given the data they were given. In computer science there's an old saying: garbage in, garbage out.

Cybersecurity: The surreal aspect of cybersecurity in 2016 is that is continues to be such a major issue. Yahoo! made the news earlier in the year when they revealed nearly 500 million customer records were hacked. But that was peanuts compared to later in the year when Yahoo! again reported over one billion customer email accounts were compromised. No wonder Verizon is looking to renegotiate their deal.

Computing systems have become so complex and internetworked that there's virtually no way to properly secure them using our current approaches. Maybe we should turn to AI instead: "Alexa, please secure my online accounts." Now that would be surreal.

# # #

ABOUT THE AUTHOR

Scott Tilley is a professor in the Department of Engineering Systems at the Florida Institute of Technology, president of the Big Data Florida user group, president of the Center for Technology & Society, and a Space Coast Writers' Guild Fellow. His recent books include *Friends* (Anthology Alliance, 2016), *Technology Hacktivists Anonymous* (CTS Press, 2016), and *Systems Analysis & Design* (Cengage, 2016). He writes the weekly "Technology Today" column for *Florida Today*. More information about his other books is available at http://www.amazon.com/author/stilley.

www.ingramcontent.com/pod-product-compliance
Lightning Source LLC
Chambersburg PA
CBHW071549080326
40690CB00056B/1615